Life-Cycle Assessment in Building and Construction:
A state-of-the-art report, 2003

Other titles from the Society of Environmental Toxicology and Chemistry (SETAC):

Life-Cycle Impact Assessment: Striving towards Best Practice
Udo de Haes, Finnveden, Goedkoop, Hauschild, Hertwich, Hofstetter, Jolliet, Klöpffer, Krewitt, Lindeijer, Müller-Wenk, Olsen, Pennington, Potting, Steen
2002

Sixth LCA Symposium for Case Studies (Presentation summaries)
1998

Fifth LCA Symposium for Case Studies (Presentation summaries)
1997

Life-Cycle Impact Assessment: The State of the Art, 2nd edition
Barnthouse, Fava, Humphreys, Hunt, Laibson, Noesen, Norris, Owens, Todd, Vigon, Weitz, Young, editors
1997

Public Policy Applications of Life-Cycle Assessment
Allen, Consoli, Davis, Fava, Warren, editors
1997

Simplifying LCA: Just a Cut?
Christiansen, editor
1997

Integrating Impact Assessment into LCA
Udo de Haes, Jensen, Klepffer, Lindfors, editors
1994

Life-Cycle Assessment Data Quality: A Conceptual Framework
Fava, Jensen, Lindfors, Pomper, De Smet, Warren, Vigon, editors
1994

A Conceptual Framework for Life-Cycle Impact Assessment
Fava, Consoli, Denison, Dickson, Mohin, Vigon, editors
1993

Guidelines for Life-Cycle Assessment: A "Code of Practice"
Consoli, Allen, Boustead, Fava, Franklin, Jensen, De Oude, Parrish, Perriman, Postlethwaite, Quay, Séguin, Vigon, editors
1993

A Technical Framework for Life-Cycle Assessment
Fava, Denison, Jones, Vigon, Curran, Selke, Barnum, editors
1991

For information about SETAC publications, including SETAC's international journal *Environmental Toxicology and Chemistry*, contact the SETAC Administrative Office nearest you.

SETAC North America
1010 North 12th Avenue
Pensacola, Florida, USA 32501-3367
T 850 469 1500
F 850 469 9778
E setac@setac.org

SETAC Europe
Avenue de la Toison d'Or 67
B-1060 Brussels, Belgium
T 32 2 772 72 81
F 32 2 770 53 83
E setac@setaceu.org

www.setac.org

Environmental Quality Through Science®

Life-Cycle Assessment in Building and Construction:
A state-of-the-art report, 2003

Edited by

Shpresa Kotaji
Global Product EHS Group
Brussels, Belgium

Agnes Schuurmans
INTRON
Sittard, The Netherlands

Suzy Edwards
Centre for Sustainable Construction
Building Research Establishment (BRE)
Brighton, United Kingdom

Coordinating Editor of SETAC Books
Andrew Green
International Lead Zinc Research Organization
Raleigh, North Carolina, USA

Published by the Society of Environmental Toxicology and Chemistry (SETAC)

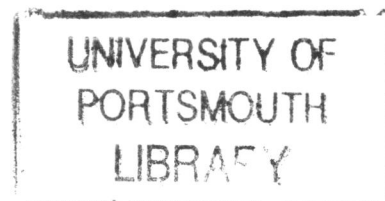

SETAC PRESS

Cover by Michael Kenney Graphic Design and Advertising and Kerri Charlton
Indexing by IRIS

Library of Congress Cataloging-in-Publication Data

Life-cycle assessment in building and construction: a state-of-the-art report, 2003 / edited by Shpresa Kotaji, Agnes Schuurmans, Suzy Edwards.
        p. cm.
      Includes bibliographical references and index.
      ISBN 1-880611-59-7 (alk. paper)
       1. Building--Planning. 2. Civil engineering--Planning. 3. Building materials--Service life. 4. Product life cycle. 5. Buildings--
Environmental aspects. 6. Environmental impact analysis. 7. Green architecture. I. Kotaji, Shpresa, 1965- II. Schuurmans, Agnes, 1968- III.
Edwards, Suzy, 1970- IV. SETAC-Europe.

TH153.L475 2003
720'.47--dc21

                              2003052625

© 2003 Society of Environmental Toxicology and Chemistry (SETAC)
SETAC Press is an imprint of the Society of Environmental Toxicology and Chemistry.
No claim is made to original U.S. Government works.

International Standard Book Number 1-880611-59-7
Printed in the United States of America
10 09 08 07 06 05 04 03   10 9 8 7 6 5 4 3 2 1

∞ The paper used in this publication meets the minimum requirements of the American National Standard for Information Sciences—Permanence of Paper for Printed Library Materials, ANSI Z39.48-1984.

Reference Listing: Kotaji S, Schuurmans A, Edwards S. 2003. Life-cycle assessment in building and construction: A state-of-the-art report, 2003. Pensacola FL, USA: Society of Environmental Toxicology and Chemistry (SETAC). 102 p.

SETAC Publications

Books published by the Society of Environmental Toxicology and Chemistry (SETAC) provide in-depth reviews and critical appraisals on scientific subjects relevant to understanding the impacts of chemicals and technology on the environment. The books explore topics reviewed and recommended by the Publications Advisory Council and approved by the SETAC North America Board of Directors and the SETAC World Council for their importance, timeliness, and contribution to multidisciplinary approaches to solving environmental problems. The diversity and breadth of subjects covered in the series reflect the wide range of disciplines encompassed by environmental toxicology, environmental chemistry, hazard and risk assessment, and life-cycle assessment. SETAC books attempt to present the reader with authoritative coverage of the literature, as well as paradigms, methodologies, and controversies; research needs; and new developments specific to the featured topics. The books are generally peer reviewed for SETAC by acknowledged experts.

SETAC Publications, which include Technical Issue Papers (TIPs), workshop summaries, newsletter (*SETAC Globe*), and journal (*Environmental Toxicology and Chemistry*), are useful to environmental scientists in research, research management, chemical manufacturing and regulation, risk assessment, life-cycle assessment, and education, as well as to students considering or preparing for careers in these areas. The publications provide information for keeping abreast of recent developments in familiar subject areas and for rapid introduction to principles and approaches in new subject areas.

SETAC recognizes and thanks the past SETAC books editors:

C.G. Ingersoll, Midwest Science Center
U.S. Geological Survey, Columbia, Missouri, USA

T.W. La Point, Institute of Applied Sciences
University of North Texas, Denton, Texas, USA

B.T. Walton, U.S. Environmental Protection Agency
Research Triangle Park, North Carolina, USA

C.H. Ward, Department of Environmental Sciences and Engineering
Rice University, Houston, Texas, USA

UNEP-SETAC Life Cycle Initiative

Established in 2000 and officially launched in 2002 by the United Nations Environment Programme (UNEP) and the Society of Environmental Toxicology and Chemistry (SETAC), the Life Cycle Initiative, through international conferences, workshops, and working groups, builds on the ISO 14040 standards and intends to establish approaches with the best practice for a Life Cycle Economy. With a mission to develop and disseminate practical tools for evaluating the opportunities, risks, and trade-offs associated with products over their entire life, the Life Cycle Initiative promotes life cycle thinking, which has significant potential for advancing environmental protection worldwide, resulting in a cleaner, safer, sustainable society. Since its inception, the Life Cycle Initiative continues to build bridges between developed and developing countries and researchers and users.

For more information, please visit the UNEP-SETAC Life Cycle Initiative website at http://www.uneptie.org/pc/sustain/lcinitiative/ or contact UNEP or SETAC:

United Nations Environment Programme
Division of Technology, Industry & Economics (DTIE)
Production and Consumption Unit
39-43 Quai André Citroën
75729 Paris, Cedex 15, France
T +33 1 4437 1450; F +33 1 4437 1474
E sc@unep.fr

SETAC North America
1010 North 12th Avenue
Pensacola, FL 32501-3367 USA
T 850 469 1500; F 850 469 9778
E setac@setac.org

SETAC Europe
Avenue de la Toison d'Or 67
B-1060 Brussels, Belgium
T 32 2 772 72 81; F 32 2 770 53 86
E setac@setaceu.org

Preface

Environmental aspects are one of many diverse properties of products. They can be studied in several ways, including use of environmental life-cycle assessment (LCA) tools. It has been recognised that the integration of LCA into daily practice could help achieve sustainable practices. This awareness has resulted in interest for LCA in many sectors of industry, including the building and construction (B/C) sector. In most European countries and North America, LCA has become a well-known instrument to study the significant environmental effects caused by B/Cs.

This specific interest for LCA application has been recognised by the Society of Environmental Toxicology and Chemistry (SETAC) Europe LCA Steering Committee, which decided to create a working group to deal with LCA specificities in B/C. The SETAC organisation functions as a platform for bringing together LCA practitioners in the B/C sector. The large number of participants from all over the world confirmed the need for such a specific platform. Besides SETAC members, members of the RILEM TC-EID/ECB committee took part in this work.

In this report, we show the important issues that arise when LCAs are performed in the B/C sector, the way they are dealt with today, and the needs for the future. The target groups are 'newcomers' in the field of LCA in B/C but with some LCA background, as well as more experienced LCA practitioners who wish to gain an insight into current practices in Europe. We hope that this report contributes to a further development of LCA in the B/C sector and to the harmonisation of all the work that is so actively being carried out at the moment.

———Shpresa Kotaji
———Suzy Edwards
———Agnes Schuurmans

Contents

List of Figures

List of Tables

About the Editors

Shpresa Kotaji studied Industrial Chemistry in Brussels, Belgium. She joined ICI Polyurethanes (now Huntsman Polyurethanes) in 1988 in the Research and Development Team, where she worked on developing CFC-free polyurethane integral skin foams and new generations of polyurethane products for the automotive industry. In 1995, she joined the International Environmental Affairs Group and became the company specialist on environmental life-cycle assessment (LCA). She has gained more experience in a wider range of environmental issues and is now part of the Global Product EHS Group and is the Issue Manager for Waste Management, Environmental and Energy Labelling, Ozone and Climate Change issues. She is also the Issue Manager for Life-Cycle Assessment and Environmental Labelling of ISOPA, the European trade association for producers of di-isocyanates.

Suzy Edwards is a Principal Consultant in the Centre for Sustainable Construction at Building Research Establishment (BRE), the United Kingdom's leading organisation for research and consultancy in construction. Edwards leads the BRE team providing environmental information to materials manufacturers and users. She is an author of the UK methodology for applying LCA to construction products and is the manager of the BRE Certification Scheme for Environmental Profiles of construction products. She is a member of ISOTC59, the technical committee that deals with international standards for building.

Agnes Schuurmans works as a senior consultant in the field of environmental aspects of building materials, environmental life cycle analyses and environmental product declarations. She has more than 10 years' experience in LCA for the building industry. She is employed by INTRON, a Dutch research and consultancy firm specialising in technical and environmental issues in the building sector. Agnes is a member of the SETAC Europe LCA Steering Committee and was convenor and secretary of the SETAC Europe Working Group LCA in Building and Construction.

Executive Summary

Increasing environmental awareness in building and construction (B/C) has resulted in the proliferation of building environmental assessment tools and application of life-cycle assessment (LCA) methodologies.

In 1998, the Society of Environmental Toxicology and Chemistry (SETAC) Europe Steering Committee created a working group to consider the application of LCA in the B/C sector. The objectives of the group were to identify important characteristics of LCA and propose guidelines or options for methodological choices as well as to propose a set of recommendations for future work. RILEM TC-EID/ECB Systematics of the Environmental Impact Databases of building materials group joined the SETAC working group.

The results of this work are reported in this state-of-the-art document. The prevailing message is the need for harmonisation of approach in order to allow LCA results from different studies to be compared and to be used to make meaningful choices in the B/C sector.

Results

The main findings of this report on the topic of LCA and B/C are summarised below. Readers are advised to refer to the glossary for definitions of the terms used.

The focus for LCA studies

The final B/C, defined by performance requirements, is the central subject of an LCA study in the B/C sector and provides the most accurate subject for any comparison.

In practice, the products or components of building or construction can provide a valid subject for the application of LCA. However, the context of the complete B/C should be reflected or at least mentioned in comparative LCAs of B/C components and incorporated whenever appropriate.

It is the designer's responsibility to design the building or construction to a suitable performance and to define the functional equivalence of alternative B/Cs. It is the responsibility of LCA practitioners (e.g., with help from manufacturers) to provide suitable building material and component combination (BMCC)–LCAs that can be brought together to the B/C level. Finally, it is the responsibility of the builder to construct the facility in accordance with industry standards.

Guidelines for different goals

If an LCA is to be meaningful, the whole life cycle of the B/C should be taken into account. The practitioner should undertake quantification of the retrospective life-cycle stages (e.g., cradle-to-gate) and assumptions about the prospective life-cycle stages (e.g., gate-to-grave) as appropriate.

If the goal is to inform users about the environmental impact of BMCCs, cradle-to-gate data are measured, and information on B/C subsequent life-cycle stages will be based on qualitative or quantitative assumptions. As a minimum, and if applicable, an indication of probable service life and end-of-life scenarios should be provided.

Information about subsequent life-cycle stages should preferably be presented separately in order to allow the next users to use different scenarios.

If the goal is alternative BMCC comparative assertion, (prospective) life-cycle stages of the B/C can be omitted if their performance at these stages is considered equivalent. As a minimum, those B/C considerations should be discussed.

If the goal is BMCC product development or process optimisation, the B/C perspective can be omitted if it is clear that the B/C (structure or performance) is not influenced by the performance of the newly developed or optimised BMCC. As a minimum, the B/C perspective should be discussed.

True functional equivalence for buildings or components can only be assessed at the level of the whole B/C over its entire life cycle. Comparisons of functionally equivalent B/Cs will be based on a set of B/C building performance characteristics.

The choice of a functional unit to compare alternative BMCCs should be defined to reflect the functional equivalence of the B/C.

Primary and secondary functions of BMCCs must be assessed in the context of the performance requirements of the B/C.

Service life scenarios

Currently, there are no commonly accepted rules for how to define B/C or BMCC service life, frequency of maintenance, and replacement. Developments in service life planning to determine actual service life and remaining value of the BMCC will be useful.

Actual service lifetimes, based on experience, are generally favoured over potential or maximum technical lifetimes.

Dynamic modelling of environment-related durability of BMCCs has been neglected to date and needs to receive more attention.

End-of-life scenarios

Ending the life cycle with the generation of waste streams or further accounting for waste treatment are both frequently applied in BMCC–LCAs and B/C–LCAs. The waste scenario can be formulated in 2 ways, depending on the study goals:
  1) the current waste management scenario, or
  2) a hypothetical waste management scenario (e.g., based on developing recycling technologies).

If the current scenario is chosen, it is also useful for designers to be able to apply future scenarios, for example, for a design-for-recycling study.

End-of-life scenarios are country dependent and sometimes building or site dependent. It is useful to have the freedom to choose the appropriate scenario as well as a common national default scenario.

Allocation

System boundaries and allocation rules differ between projects and countries. Although there is an international knowledge transfer and growing consensus, a further harmonisation is desired for information transfer of LCAs between studies and countries.

Multi-input and multi-output processes

It is the responsibility of the LCA practitioner to use the appropriate allocation principle for products. The allocation method should reflect the causality of material and energy flows.

End-of-life allocation

There are different views on how to deal with allocation for long-life applications such as building products and B/C. Main discussion points include the following:

- The applicability of system expansion to avoid allocation and to apply subtraction methods, especially with regard to the comparability of BMCC-DLCAs for combination in B/C-DLCAs.
- The applicability of subtraction methods for (semi-)closed loop recycling processes, also especially for BMCC-DLCAs to be combined in B/C-DLCAs.
- The percentage of recycled material input to allocate to the current product that is being produced. There are 2 main views:
    1) allocate the current percentage of recycled material only, or
    2) allocate the future percentage of material that is going to be recycled.
- The fact that allocation procedures concentrating on the devaluation ('quality loss') of a material over a life cycle can be more appropriate than using factors on recycling rates.

Although this discussion is not building specific (it is relevant for all long-life products), it is essential for LCA of B/C because the use of BMCC-DLCAs in B/C-DLCAs for comparative purposes is only possible when comparable allocation principles are applied. Ideally, LCA data on end-of-life processes should be provided in a non-allocated format to allow decision-makers to choose the appropriate allocation method for the goal of the study.

LCA and indoor air quality

Hazardous substances are not easily dealt with in B/C-DLCAs in BMCC-DLCAs. Under certain conditions, hazardous substances can be released from materials or formed when some materials combine, and these impact building indoor air quality (IAQ) or the factory working environment and result in health issues. IAQ is highly linked to building design.

The relation between LCA and the working environment is being dealt with in another SETAC working group.

International differences

At the moment, declaration and labelling systems for BMCC-DLCAs are very difficult or impossible to exchange between actors and countries. Further discussions, development, and harmonisation are essential for a successful use of LCA in the B/C sector.

Recommendations

Further research for LCAs in B/C is required in the following areas:
- Harmonisation of BMCC–LCAs that will be used for B/C–LCAs. There should be a particular focus on harmonisation of system boundaries, modelling of maintenance, definition of end-of-life scenarios and allocation.

Related areas for research, either ongoing or needing further attention, that will impact upon the application of LCA to B/C include
- the relationship between durability and environmental effects (dynamic modelling),
- the service life of BMCCs and service life of the B/C and the development of consistent scenarios,
- the clarification of the interrelation between LCA and energy environmental impact assessment (EIA),
- the ability of LCA to deal with IAQ and hazardous substances, and
- the combination of LCA with life-cycle costing (LCC).

Knowledge transfer and communication are required further to this state-of-the-art report, especially in the field of product declarations and labelling for BMCCs and B/C.

1 Introduction

Why Life-Cycle Assessment in Building and Construction?

The development of life-cycle assessment (LCA) in the building and construction (B/C) sector is accelerating. Much effort, both from researchers and from the building industry, government, or designers has been, or is being, directed into LCA projects worldwide. The importance of environment-related product information by means of LCA is broadly recognised, and LCA is considered one of the tools to help achieve sustainable building practices.

Applying LCA in the B/C sector has become a distinct working area within LCA practice. This is not only due to the complexity of B/Cs but also because of the following factors, which combine to make this sector unique in comparison to other complex products.

Buildings and constructions have extremely long lifetimes, often more than 50 years. It is difficult to predict the life cycle 'from cradle-to-grave'.

During this life span, the building or construction may undergo many changes in its form and function. These changes can be as significant, or even more significant, than the original construction. The ease with which changes can be made and the opportunity to minimise the environmental effects of changes are partly functions of the original design.

Many of the environmental impacts of a building occur during its use (energy and water consumption). Proper design and material selection are critical to minimise those in-use environmental loads.

There are many actors in the B/C column. The designer, who makes the decisions about the final building or construction or its required performance, does not produce the components, nor does he or she build the B/C. In many cases, very formal relations (e.g., obligatory open tenders) exist among commissioners, designers, and contractors.

Traditionally, each building or construction is unique and is designed as such. There is very little standardisation in whole building design. New choices have to be made for each specific situation.

The comparability of LCAs of distinct products and the way these LCAs are applied to design and construct environmentally sound B/Cs is a main point of attention in LCA practice. Several initiatives for harmonisation and standardisation of methodological developments and LCA practice in the building industry have taken place at a national level, but in general much scope remains for wider involvement and co-operation.

Life-Cycle Assessment in Building and Construction. Shpresa Kotaji et al., editors.
©2003 Society of Environmental Toxicology and Chemistry (SETAC). ISBN 1-880661-59-7

Another point worth mentioning is the distinction between LCA practitioners and other actors who often know little about LCA. In many cases, the LCA practitioners tend to work at the level of individual materials and products, while the user of LCA data (i.e., designers, etc.) are concerned with the whole B/C performances. The challenge is to make sure that the principles of LCA are not violated and that the LCA data on materials and products can be brought together for or by people with limited LCA knowledge. LCA results have to be made accessible to decision-makers.

The Society of Environmental Toxicology and Chemistry (SETAC) Europe Working Group LCA in Building and Construction aims to bring initiatives and ideas from all over the world together in a state-of-the-art report for LCA in the B/C sector.

SETAC Working Group

The SETAC Europe LCA Steering Committee founded the SETAC Working Group in 1998. The members of RILEM TC-EID/ECB Systematics of the Environmental Impact Databases of building materials took part in the Working Group. (RILEM is the International Association for Building Materials and structures, a network of technical experts advancing the knowledge of building materials and structures.) A full membership list is included in Appendix A. Table 1-1 shows the members who contributed actively to the work in this report.

The work described in this report has a relationship with activity being carried out in other SETAC working groups. For details of these reports, see the SETAC web site (www.setac.org). Where relationships exist, they will be mentioned in the text of this report.

Objectives and Target Groups

The objectives of the working group are defined as follows:
- to identify important characteristics of LCA in B/C,
- to identify LCA methodological subjects that are relevant for B/C and that are not covered by current methodology, and
- to elaborate identified topics into guidelines or options for methodological choices.

The results are reported in this state-of-the-art document. The prevailing message is one of need for harmonisation of approach in order to allow LCA results from different studies to be compared and to be used to make meaningful choices in the B/C sector. If it is relevant for a study, however, the LCA practitioner can go beyond this state of the art or may choose another approach.

The main target groups of this state-of-the-art report are
- newcomers to the field of LCA in B/C,
- LCA practitioners who want to learn more about current practice in different European countries (system limits, allocation rules) and how to build up LCA for a building,
- LCA researchers who need to identify topics for further research, and
- producers of LCA tools in the building field.

Readers should have a sound knowledge of LCA because this report does not explain the basics of the procedures but discusses sector-specific particular methodological issues. Examples are provided to further explain issues. These examples should facilitate the reading of this report by non-LCA experts, for example, representatives of industry and those who wish to commission LCAs.

Table 1-1 Members who contributed actively to the work in *Life-Cycle Assessment in Building and Construction*

Organisation	Country	Member
Athena Sustainable Materials Institute	Canada	Wayne Trusty
Building Research Establishment (BRE)	UK	Suzy Edwards
Carillion Building	UK	Andrew Horsley
CSTB	France	Jean François le Teno
CSTC-BBRI-WTCB	Belgium	Jan Desmyter
Eidgenössische Technische Hochschule (ETH) Zürich / Basler & Hofmann	Switzerland	Annick Lalive d'Epinay
Huntsman	Belgium	Shpresa Kotaji
INTRON	Netherlands	Agnes Schuurmans
KTH / Ingenieurbüro Trinius	Sweden, Germany	Wolfram Trinius
Randa	Spain	Marta Vallès
Rijkswaterstaat (Dutch Ministry of Transport, Public Works and Water Management)	Netherlands	Joris Broers
SBI (together with DTI)	Denmark	Hanne Krogh Klaus Hansen
TNO-Bouw	Netherlands	Adrie de Groot-Van Dam
UBA Berlin	Germany	Hans-Hermann Eggers
University of Brighton	UK	Andrew Miller
University of Stuttgart-IKP	Germany	Johannes Kreissig
VTT	Finland	Tarja Häkkinen

This Report

After a general elaboration of the subject, methodological issues are worked out according to the LCA steps of SETAC's *Guidelines for Life-Cycle Assessment: A 'Code of Practice'* (Consoli et al. 1993) and the International Organization for Standardization 14040 series (ISO 1997). The contents are as follows:

- Chapter 2 describes the LCA application in B/C: the products to be analysed and the steps in the LCA process.
- Chapter 3 deals with the goal and scope definition: possible goals and applications, functional unit, and other goal and scope issues.
- Chapter 4 provides information about the inventory process.

- Chapter 5 describes topics of the impact assessment that are specific for the building industry.
- Chapter 6 brings attention to sensitivity analysis issues.
- Chapter 7 discusses the presentation and communication requirements.
- Chapter 8 links LCA to 2 life-cycle and environmental instruments: life-cycle costing (LCC) and environmental impact assessment (EIA), respectively.
- Chapter 9 provides an outlook for the future, as defined by the SETAC Working Group.

Appendix A contains the full member list of the Working Group. Appendixes B and C provide background information for topics dealt with in this report. Appendix D contains an overview of LCA studies in the B/C sector.

For ease of reading, we will use the abbreviation 'B/C' to refer to building and/or construction and 'BMCC' to refer to building materials, components, or combinations of these into building elements.

References

Consoli F, Allen D, Boustead I, Fava J, Franklin W, Jensen A, de Oude N, Parrish, R, Perriman R, Postlethwaite D, Quay B, Séguin J, Vigon B. 1993. Guidelines for Life-Cycle Assessment: A 'Code of Practice'. Pensacola FL, USA: Society of Environmental Toxicology and Chemistry (SETAC).

[ISO] International Organization for Standardization. 1997. ISO 14040: Environmental management—Life cycle assessment—Principles and framework. Geneva, CH: ISO.

2 Framework for LCA in Building and Construction

Product to Be Analysed: Performance Requirements of a Building or Construction

The 'product' under study in the life-cycle assessment (LCA) of a building or construction (B/C) is the building or construction itself, defined according to a certain level of performance and including all the necessary material processes. When parts of a B/C are studied, their functional equivalence within the final B/C is essential. The B/C product consists of different components—building materials, products or components—that we will refer to as 'building material and component combinations' (BMCCs) (Llewellyn and Edwards 1997). We will come back to this issue in Chapter 3 when we deal with the functional unit.

During the design of a B/C, environmental determinants are only one of the determining aspects besides, amongst others, cost, aesthetics, and technical, functional, or legal requirements (Lampugnani 1995). It is the aim of the architect to integrate all these determinants in an optimal way in the design to achieve the required B/C performance. LCA should give reliable information concerning the environmental aspects of the B/C. The B/C performance concept and interaction with B/C determinants is shown in Figure 2-1.

There are always several possible alternative solutions, which can bring about the same required performance and which have to be compared regarding their different costs, environmental consequences, or any other determining aspect.

For constructions, such as dikes, the environmental performance of the constituent material as well as the construction impact on landscape and biodiversity will often dominate the construction life-cycle environmental impacts.

For buildings, such as dwellings or warehouses, life-cycle environmental impacts are often dominated by energy consumption, in space heating or lighting, during the use phase. It has been estimated that the use phase in conventional buildings represents approximately 8% to 90% of the life-cycle energy use, while 10% to 20% is consumed by the material extraction and production and less than 1% through end-of-life treatments. With the development of energy-efficient buildings and the use of less-polluting energy sources, the contribution of the material production and end-of-life phases is expected to increase in the future. Materials and end-of-life environmental performances already dominate impacts that are not energy-related, such as emissions to water.

It is important to note at this stage that the building location and orientation will have consid-

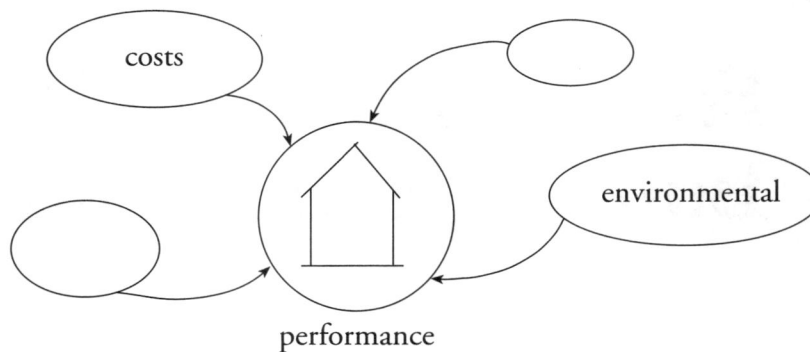

Figure 2-1 The performance concept for buildings and constructions

erable impacts on the building energy consumption and therefore on the overall environmental
impacts even if the same BMCCs and construction
techniques are used. For example, the benefits
from the use of passive solar energy or natural
ventilation will need to be incorporated. The B/C–
LCA is more than the addition of BMCC–LCAs.
Designing a building with low environmental loads
requires the matching of materials and products,
regardless of their impacts at the material or product level, to the specific design and site to optimise
the overall building environmental impact.

Estimations of the environmental impact of
the whole-building stock show that the relation
between impacts due to use and impacts due to
construction activities are very different, according
to which type of impact is considered, and indicate

that generalisation is hardly possible (Kohler et al.
1999).

Beyond the building

A building or construction functions in certain
surroundings, and the infrastructure it is part of
can be considered the final product, instead of the
B/C. There are initiatives to extend LCA to this
field, at the moment mainly for settlements of
dwellings and offices (BRE 2000, 2001; DGMR
and Stichting Sureac 2000) but in the future probably also for constructions such as roads or waterworks (Llewellyn and Edwards 1997; Kohler et al.
1999). This report focuses only on the B/C itself.
Infrastructure demands are regarded as part of the
performance requirements.

The final building or construction, defined by performance requirements, is the central subject
of an LCA study in the B/C sector and provides the most accurate subject for any comparison. In
practice, the products or components of a building or construction can provide a valid subject for
the application of LCA. However, the context of the complete building or construction shall be
reflected or at least mentioned in comparative LCAs of building or construction components and
incorporated whenever appropriate (see 'Functional Units,' Chapter 3).

_hi_ ☺
LCAs of BMCCs

$2012!$

```
┌──────────────────────────────┐        ┌──────────────────────────┐
│      Product phase            │  ⟵    │      BMCC–LCAs            │
│ Materials and products in     │        │    cradle-to-grave       │
│ the design ('as you buy')     │        │                          │
└──────────────────────────────┘        └──────────────────────────┘
              │
              ▼
┌──────────────────────────────┐        ┌──────────────────────────┐
│ Building and construction     │  ⟵    │ BMCC–LCAs: transport,    │
│          phase                │        │ construction processes,  │
│ Transport, construction, use, │        │ maintenance, demolition  │
│ maintenance, demolition       │        ├──────────────────────────┤
│                               │  ⟶    │ LCA data on use of the   │
│                               │        │        building          │
└──────────────────────────────┘        └──────────────────────────┘
              │
              ▼
┌──────────────────────────────┐        ┌──────────────────────────┐
│    Waste treatment phase      │  ⟵    │      BMCC–LCAs           │
│ Waste treatment of materials  │        │    waste treatment       │
│     and products              │        │                          │
└──────────────────────────────┘        └──────────────────────────┘
```

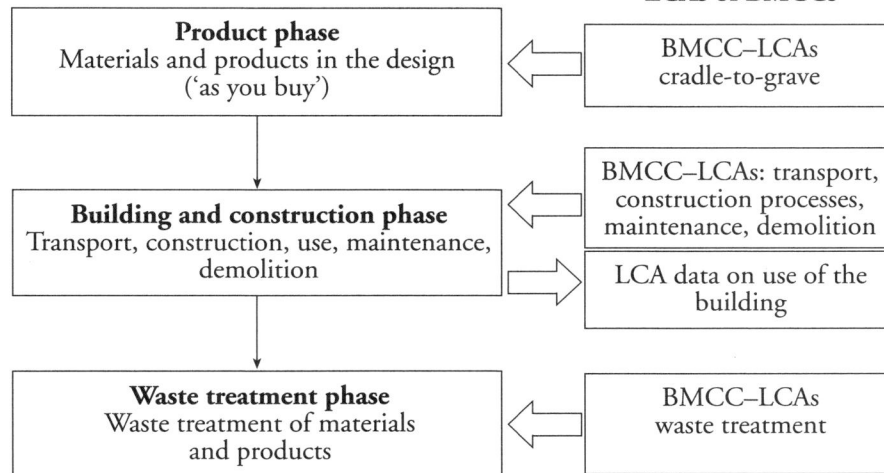

Figure 2-2 Relationship between LCA of whole buildings and BMCCs

Steps in the LCA of Buildings and Constructions

The first step is to define the goal and scope of the B/C–LCA study. The general framework for LCA in B/C (the goal and scope is further elaborated in Chapter 3) involves the following goals and LCA steps (see also Figure 2-2):

1) The life cycle of the B/C is described. What is included in the study will depend on the scope. It may include how the B/C is constructed, used, maintained, and demolished and what happens to the waste materials after demolition. These are processes that contribute to the life-cycle performance of a B/C, but which will not necessarily be included in all studies.

2) The B/C is 'broken down' to the BMCC level. This is the composition of the B/C to be analysed. The way in which the BMCCs are defined is not necessarily important; what matters is that the B/C is completely described through the addition of the BMCCs.

3) For each BMCC, the LCA of the production process (cradle-to-gate) is carried out. Their LCAs may also include the transport processes to the B/C site, the construction processes, the use and maintenance processes, the demolition processes, and the waste treatment processes for each of the waste materials defined in the B/C model. This would be a cradle-to-gate analysis.

Because the focus in the LCA shifts from product orientation in a cradle-to-gate LCA to a component and B/C orientation when construction processes, maintenance, and end-of-life are added, these stages ideally should not be aggregated as to allow other scenarios to be performed in the B/C model. When a cradle-to-gate LCA is performed, assumptions made regarding use in the B/C context may differ from the knowledge of a designer. Keeping the assumptions distinct from each other allows the designer to modify them to match his or her own requirements. Such an approach may not always be feasible, however, when a user-friendly interpretation of LCA data is required.

The BMCC–LCA results are added together, resulting in the LCA of the B/C (either life-cycle inventories [LCIs] or life-cycle impact assessment [LCIA] results). The various BMCC-LCAs should be carried out consistently according to the goal and scope.

Steps 1 and 2 require a high degree of knowledge of B/C (mainly, the working field of architects). As for Step 3, knowledge of and expertise in LCA and in the BMCC that is the subject of the LCA (knowledge of the producer, etc.) are required. It is the designer's responsibility to design the B/C in an appropriate way.

The latter requires highly standardised procedures for LCA. The ISO 14040 standards allow a wide variation in detailed approaches. Therefore, more standardised procedures must be identified beyond the ISO standard and agreed upon for application by LCA practitioners and BMCC producers in the B/C sector.

Important Issues for LCA of Buildings and Constructions

The following issues are identified as being particularly important for LCAs of B/C:

1) Goal and scope definition
   - Variety of different actors who use LCA in different applications
   - Functional unit, from the perspective of the B/C performance concept
   - Description of the B/C.

2) Inventory analysis
   - B/C life-cycle description and system boundaries (service life, dynamic or static view of the building's physical performance, etc.)
   - System boundaries, especially of the BMCC-LCAs that have to be combined
   - Allocation for long-living products such as B/C, for example, to potential re-use or recycling in the future
   - End-of-life information that is associated to the BMCC (not identical with allocation).

3) Impact assessment
   - Role of chemical substances in relation to indoor climate
   - Land use and deterioration of eco-systems (relevant because of the large quantities of minerals used in the building industry).

   (Labour issues in the construction phase [working environment] are not dealt with because this relatively new subject is already being studied by another SETAC working group.)

4) Use and application of results
   - Presentation and communication instruments that are currently being used for the applications and target groups defined in the goal and scope definition.

5) Relation to other tools
   - Life-cycle costing (LCC), being an instrument with increasing importance for architects and with a strong methodological relationship with LCA
   - Environmental impact assessment (EIA), because of the importance of local burdens for B/C and the importance of infrastructure.

Availability of life-cycle inventory (LCI) data is an important issue for all LCAs. A reference list as

It is the responsibility of LCA practitioners (with the help of producers, etc.) to provide suitable BMCC–LCA data and to make sure that the data can be brought together at the B/C level in a proper way.

provided by the members of the working group, can be found in Appendix D.

Currently available software tools for LCAs of B/C are also listed in Appendix D.

Other generic LCA issues, for example, representativity and data collection, are not specific to B/C and are therefore not examined in this report.

References

[BRE] Building Research Establishment. 2001. Checklist for sustainable settlements. Department of the Environment, Transport and the Regions (DETR) research. Garston, Watford, UK: BRE.

[BRE] Building Research Establishment. 2002. Sustainability indicators for infrastructure. Department of the Environment, Transport and the Regions (DETR) research. Garston, Watford, UK: BRE.

DGMR and Stichting Sureac. 2000. Greencalc [computer program]. NL.

IVAM Environmental Research. 2001. Eco-Quantum on location, experimental. Amsterdam, NL. Forthcoming.

Kohler N, Hassler U, Paschen H. 1999. Stoffströme und Kosten im Bereich Bauen und Wohnen. Berlin, DE: Studie im Auftrag der Enquete Kommission zum Schutz von Mensch und Umwelt des deutschen Bundestages.

Kohler N, Schwaiger B. 1998. Sustainable management of buildings and building stocks. Green Building Conference. Vancouver BC, CAN: International Council for Research and Innovation in Building and Construction (CIB).

Lampugnani VM. 1995. Die Modernität des Dauerhaften, Essays zu Stadt, Architektur und Design. Aus dem Italienischen von Moshe Kahn, 51. Band der Reihe Kleine Kulturwissenschaftliche Bibliothek. Berlin, DE: Verlag Klaus Wagenbach.

Llewellyn J, Edwards S, editors. 1997. Towards a framework for environmental assessment of building materials and components. Brite Euram 7890 report. Watford, UK: Building Research Establishment (BRE).

3 Goal and Scope Definition

Goals, Applications, Target Groups

In this section, we analyse the goals or, in other words, the applications and the target groups of life-cycle assessment (LCA) in building and construction (B/C). We distinguish 2 levels:

1) Building materials component combination (BMCC)–LCAs: On this level, LCAs are undertaken for materials, products, services, etc.
2) Whole B/Cs: On this level, LCA is conducted for B/Cs. BMCC–LCAs (either life-cycle inventory [LCI] or life-cycle impact assessment [LCIA]) are used to create the LCA of the whole B/C.

The level chosen for the LCA will be determined by the goal. First, we explain the significance of position in the life cycle and corresponding knowledge and responsibility; then, we describe the potential goals and target groups for each level.

Position in the life cycle

In an LCA, on both B/C and BMCC levels, the whole life cycle of the B/C is the central subject. Part of the life cycle can be exactly known (the past), part of it can be influenced (the 'now'), and part of it can only be predicted (the future). This is shown in Figure 3-1.

In general, we can say that the future is always situation specific (site and time related). The more we look to the future, the less precise we can be in our LCA, and the LCA becomes 'prospective' in the context of B/C. (The SETAC Working Group on Scenario Development defines a prospective LCA as an LCA of future product systems.) LCA practitioners cannot be fully responsible for future life-cycle stages; they can only predict on the basis of their experience and develop scenarios accordingly.

An LCA practitioner for B/C should be aware that data are needed from the past (from BMCC–LCAs, e.g.), from the 'now' (to be defined by the designer), and from the future (to be modelled with input from the designer, commissioner, or user).

LCA for B/Cs

The goal on the level of B/Cs will usually be to improve the design of B/C to minimise the environmental impact over the whole life cycle.

Benefits that may result from this include
- reduction of environmental impacts,
- reduction of life-cycle costs,

Past (certain)	Now (to be influenced)	Future (scenarios)	Time related
			Prospective
Material production	Assembly	Use and end of life	Site related

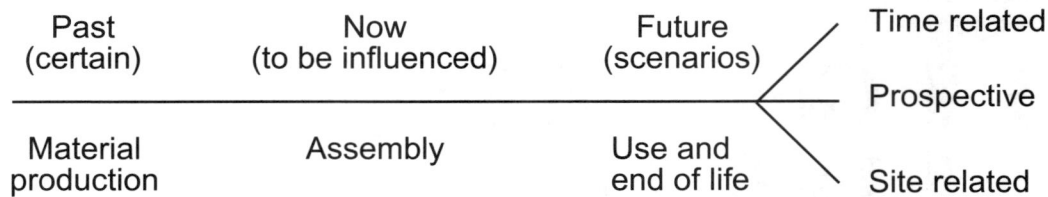

Figure 3-1 Knowledge certainty in the life cycle of a building or construction

- reduction of lifetime environmental risk, and
- better marketing opportunities for the B/C.

Target groups are designers or commissioners of a B/C and their clients (developer, investor, owner, occupier, and tenant).

Policy-makers may also be initiators of a B/C–LCA. Their goals may include minimisation of impacts from national building stocks, promotion of environmental buildings, or investigation of the impacts of subsidies, taxes, regulation on environmental performance, etc.

LCA for BMCCs

The initiator of an LCA of a material, product, component, or service may be the producer or the service provider, but other initiators are also possible (e.g., government or major client). The LCA goals can be internally or externally oriented.

Typical internal applications include
- identifying opportunities to improve the environmental aspects of the production of the product,
- decision-making (e.g., strategic planning, priority setting, product or process [re]design), and
- life-cycle management (there is a SETAC Working Group on Life-Cycle Management).

For internal LCAs, when there is no intention to compare products, it is not always necessary to take the whole life cycle of a B/C into account. In fact, such LCAs are not BMCC–LCAs, as described before, because they do not have to fulfil the modular requirement of being incorporable

into B/C LCAs. Internal improvement use of LCA is not specific for B/C and is therefore not further dealt with in this report.

For external (often public) use, the following goals, target groups, and subsequent information responsibilities can be distinguished:

1) Transferring information to the next actor in the product chain, with the final aim to provide the market with all BMCC–LCAs required for the whole life cycle of a B/C:
   - product comparisons to position the product (target group: clients),
   - policy-making (e.g., promotion of secondary materials, taxes), and
   - benchmarking.

2) Transferring information to the next actor in the product chain.
   The next actor can be a next producer or the designer or specifier, etc. Data on BMCCs can be from cradle-to-gate, to the point of construction, that is, as they are installed in the B/C, or to the end of their lives. The data to the point of installation in the B/C will be 'certain'. Data to the end-of-life will be based on predictions of what is most probable in 'typical' circumstances. Which life-cycle stages are taken into account will depend on the goal of the study, the particular knowledge of the intended life for a B/C, for example, the expected lifetime or the maintenance, and the interests of the LCA practitioner. It is preferable that assumptions or predictions about the expected life cycle of a BMCC (at the elemental level, e.g., wall type, floor construction) are provided in a desegregated way, that is, that

data to the point of construction should be provided as well as 'typical' lifetime data. The data user can then decide on what is most appropriate to their specific circumstances.

3) Product comparisons to position the product (target group: clients)

Used either as product declarations or comparative assertions, this type of LCA is mainly a marketing instrument. The result of the comparison can be used to make product choices or for drawing up guidelines for architects. In this case, the B/C perspective must be taken into account in a fully quantitative way, unless the life-cycle stages are similar under the described conditions.

Where products are identical, for example, two engineering bricks of the same specification, manufactured by different companies, any product comparisons are not building specific and therefore not dealt with in this report.

Product comparisons based on fully equivalent functionality do need to be brought into the building context in order to make a valid comparison and these comparisons are dealt with in this report.

4) Policy-making (e.g., promotion of secondary materials, taxes)

Policy-making is not dealt with in this report because it is not building specific, although policies can have an impact on the LCA of B/C or BMCCs, too. If, for example, policy promotes secondary aggregates to be used in concrete, this may (in some cases) influence the cement content and thus the LCA.

5) Benchmarking

Benchmarking also is not building specific and is not dealt with in this report.

To undertake a meaningful LCA, the whole life cycle of the B/C should be taken into account. The practitioner should undertake quantification of the retrospective life-cycle stages (e.g., cradle-to-gate) and make qualitative, semiquantitative, or quantitative assumptions about the prospective life-cycle stages (e.g., gate-to-grave) as appropriate:

- If the goal is to inform users about the environmental impact of BMCCs, cradle-to-gate data are quantitatively measured, and information on B/C subsequent life-cycle stages will be based on qualitative or quantitative assumptions. As a minimum, and if applicable, an indication of probable service life and end-of-life scenario should be provided.

Information about subsequent life-cycle stages preferably should be presented separately, in order to allow the next users to use different scenarios.

- If the goal is alternative BMCC comparative assertion, (prospective) life-cycle stages of the B/C can be omitted if their performance at these stages is considered equivalent. As a minimum, those B/C considerations should be discussed.
- If the goal is BMCC product development or process optimisation, the B/C perspective can be omitted if it is clear that the B/C (structure or performance) is not influenced by the performance of the newly developed or optimised BMCC. As a minimum, the B/C perspective should be discussed.

Functional Units

B/C functional units

True functional equivalence for B/Cs functions can be assessed only at the level of the whole structure over its entire life cycle, defined according to a series of preestablished building performance characteristics. Examples of building performance characteristics are presented in Table 3-1, and the choice will depend on the study goals. Other bases of comparison for buildings will include criteria such as m^2 inner space, m^2 building, m^3, number of inhabitants, etc.

Attempts are being made to standardise the functional units for buildings, for example, in a Dutch project (Stichting NVTB Projecten 2000), but results are not yet available. In a European LCA project (Llewellyn and Edwards 1997b), it was concluded that it is better to rely on the designer's expertise for the identification of appropriate alternatives.

BMCC functional units

Building materials and components combination LCAs are established using a functional unit that is part of a total B/C functional unit (e.g., 'per tonne' unit for a material such as mortar, 'per installed' unit such as 1 m^2 wall with R-value 2.5 m^2 K/W. In the Dutch MRPI project, functional unit is called 'analysis unit' [Stichting NVTB Projecten 2000]). Comparisons of BMCCs should be based on equal (performance) properties, such as those defined in product standards, which are appropriate, if products in one group are looked at.

Attempts have been made to standardise the functional unit for BMCCs, for example, in environmental handbooks for choosing environmentally sound building products (Anink et al. 1997; Llewellyn and Edwards 1997a). However, the European Brite Euram project, *Towards a Framework for Environmental Assessment of Buildings* (Llewellyn and Edwards 1997b), concluded that there is no commonly agreed set of unambiguous

---

Table 3-1 Examples of performance characteristics that can be used to define the functional unit of a building or construction

Examples of performance characteristics	
A1 Conformity	A5 Adaptability
A1.1 Core processes	A5.1 Adaptability in design and use
A1.2 Supporting processes	A5.2 Space systems and pathways
A1.3 Corporate image	
A1.4 Accessibility	
A2 Location	A6 Safety
A2.1 Site characteristics	A6.1 Structural safety
A2.2 Transportation	A6.2 Fire safety
A2.3 Services	A6.3 Safety in use
A2.4 Loadings to immediate surroundings	A6.4 Intrusion safety
	A6.5 Natural catastrophes
A3 Indoor conditions	A7 Comfort
A3.1 Indoor climate	
A3.2 Acoustics	
A3.3 Illumination	
A4 Service life and deterioration risks	A8 Others
A4.1 Service life (years) (technical, aesthetical, experience, etc.)	A8.1 m^2 of .. (build area, living area rooms, etc.)
A4.2 Deterioration risks	A8.2 m^3 of .. (capacity, etc.)
	A8.3 (potential) inhabitants
	A8.4 (potential) number of households

rules available yet regarding the formulation of the functions and relative performance levels of a given BMCC. Therefore, as for whole buildings, it was concluded that it is better to rely on the designer's expertise.

BMCC–LCAs using partially described functional units must be interpreted and applied carefully, unless it is clear that the building's function is not influenced by the compared BMCC alternatives. We can only be certain of functional equivalence at the BMCC level if we can be assured of the implications for equivalence throughout the rest of the building and its life. In reality, compromise is usually required, and the informed user must interpret the comparisons. Many LCA studies have been carried out on BMCCs and have been subjected to comparative assertions.

Example 1
When comparing alternative wall elements of, for example, same load bearing and same thermal properties, one should not stop at comparing the environmental profiles of those wall elements. To assess the overall impact of one wall compared to another, the study should also include, for example, the environmental load associated with more

or less concrete foundation required to achieve the required building stability and durability or the impact on the living space area which could be used as a functional requirement at the building level. (See Figure 3-2).

Example 2
When comparing 2 window frames of, for example, the same size and thermal properties, comparative assertion can be made because it is assumed that the window frame choice will not affect the building structural requirements.

Secondary functions

In B/C, the function of a BMCC is rarely unidimensional. Besides the main function (e.g., load bearing), one or several additional functions add value to it (e.g., thermal insulation or sound insulation) and, of course, to the B/C's overall performance. It is important to recognise them and in some situations, it may be appropriate to check their relevance to the building functional equivalence in a sensitivity analysis. In most cases, however, the same principle applies as for the main functions, and the designer will choose the alterna-

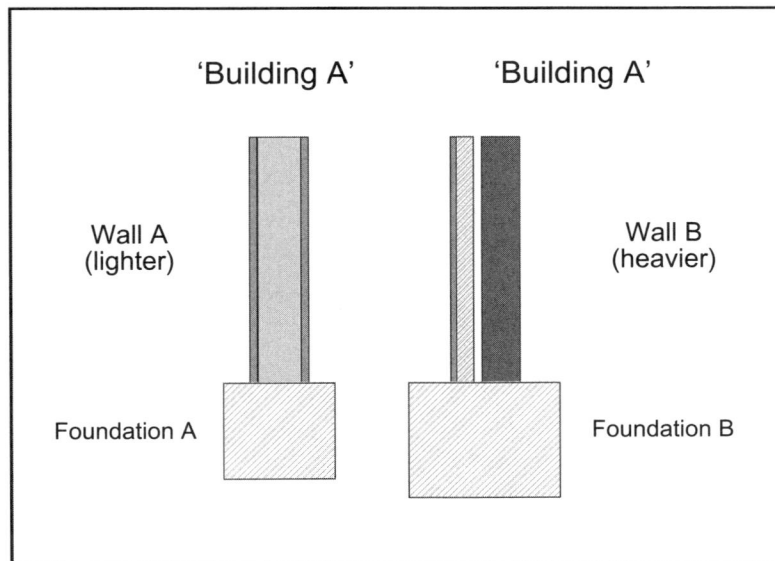

Figure 3-2 Example of the implication of wall structures for other building structure (foundation) requirements and final building LCA performances

True functional equivalence for B/Cs or BMCCs can be assessed only at the level of the whole B/C over its entire life cycle. Comparisons of functional equivalent B/Cs will be based on a set of B/C building performance characteristics. The choice of a functional unit to compare alternative BMCCs should be defined in relation to the functional performance of the B/C. Primary and secondary functions of BMCCs must be assessed in the context of the performance requirements of the B/C.

tives to be compared aware of both the primary and secondary functional equivalence.

Description of a Building

Several models have been developed to describe the life-cycle impacts of a building to inform the designer and their client of the environmental consequences of their designs. The website of the International Energy Agency Annex 31 (IEA 2001) provides a list of interesting LCA tools for B/C (up to now, mainly building tools have been developed; some construction tools are in development in the Netherlands, e.g.). New tools have emerged since then. A non-exhaustive list of some LCA software tools available includes Envest, Eco-Quantum, Greencalc, LCA HOUSE, and Athena. References can be found in Appendix D. These tools are complementary to concepts such as the BREEAM building label, Green Building Challenge, and GBA-tool as aids towards more sustainable building. Other tools such as BEES compare whole-life costs to the environmental impacts of BMCCs but do not currently include the whole building.

Because the models for designing a B/C and the way of describing components (models such as the SfB-code in the Netherlands or the UNIFOR-MAT in the U.S. [ASTM 1993]) are not normalised, it may not be surprising that the building models in these LCA tools are not similar. However, most of them require the user to select the building components close to those they are de-

signing. The basis for modelling the operational impacts of the building varies. Progress towards a common understanding of the tools is being made through European Thematic Network projects such as PRESCO (http://jbase208.eunet.be/)

A common theme in the software is the focus on the main building components, which may neglect, for example, the installations or the finishing elements. From the perspective of hitting the 'big wins', this may be considered appropriate. However, the tools are likely to expand to include more detailed aspects of the building in the future and to make sure all critical elements are recognised. This is important because components such as the finishing elements have been shown to be insignificant (Lalive d'Epinay 2000) because of their toxic substance content, high-embodied impacts, and high rate of replacement, even if their initial mass is not high. It is useful to be able to show designers where effort can be best prioritised. This has been incorporated in *The Green Guide to Specification for Housing in the UK* (Anderson and Howard 2000), in which the proportional impact of each set of functionally equivalent BMCCs, for example, 'internal partitions', has been indicated on a pie chart. Such information goes some way towards bridging the gap between BMCC-based guidance (Anink et al. 1997) and whole-building tools.

This issue is essential for 4 main reasons:
1) It is a lot of work to describe a building, and several types of documents are needed (plans,

specifications, energy calculations, details, etc.); in practice this work will never be done only for LCA.

2) If the building is not described in a neutral way, based essentially on functional decomposition (window, 1 m² of wall, etc.), it is not possible to appreciate the interrelations between performances (comfort, environmental impact, costs).

3) The professional debate is centred on the possibility of computer-readable, life-cycle–oriented descriptions of building products (product modelling). LCA is an important aspect but is not discussed for the moment.

4) The compatibility of different types of software (CAAD, databases, simulation programs, etc.) is vital for web-based collaboration. One important initiative is the Industry Foundation Classes (IFC).

The building description issue will be decisive for the introduction of integrated life-cycle analysis.

References

Anderson J, Howard N. 2000. The green guide to specification of housing. London, GB: Building Research Establishment (BRE).

Anink D, Boonstra C, Mak J. 1996. Handbook of Sustainable Building. London, GB: James & James. ISBN 1-873936-38-9.

[ASTM] American Society for Testing and Materials. 1993. Standard classification for building elements and related sitework – UNIFORMAT II. West Conshohocken PA, USA: ASTM. ASTM E1557–93.

[IEA] International Energy Agency 2001. Existing instruments, tools and databases. Annex 31. Available at www.uni-weimar.de/SCC/PRO/. Accessed 15 Mar 2003.

Lalive d'Epinay A. 2000. Die Umweltverträglichkeit als eine Determinante des architektonischen Entwurfs. Diss ETH Nr. 13610. Abteilung für Umweltnaturwissenschaften, Laboratorium für technische Chemie, Gruppe für Sicherheit und Umweltschutz. Zürich, CH: Eidgennössische Technische Hochschule (ETH).

Llewellyn J, Edwards S. 1997. Assessment of building materials and components. London, UK: CRC Ltd. ISBN 1 86081 252 X. (CRC@Construct.emap.co.uk).

Llewellyn J, Edwards S, editors. 1997. Towards a framework for environmental assessment of building materials and components. Brite Euram 7890 report. Watford, GB: Building Research Establishment (BRE).

Stichting NVTB Projecten. 2000. Manual and background document for compiling environmentally relevant product information (MRPI), SNP-R98002 version 1.2. Driebergen, NL.

4 | Inventory Analysis

The Building Life Cycle

The life cycle of a building or construction (B/C) consists of many processes. Depending on the goal of the LCA, processes can be aggregated in several life-cycle stages. Various distinctions are used, all resulting in the same total life cycle. One example uses 3 stages: production of building material and component combinations [BMCCs], use phase of the building, and waste treatment. Another uses 5 stages: production, construction, use and maintenance, demolition, and waste treatment. A guideline for the stages is not useful because the definition is related to the goal of the study. However, the following issues have to be kept in mind:

- If the life cycle of a building is divided into stages, the total of the stages should reflect the total life cycle.
- In practice, the life-cycle assessment (LCA) of a building often uses ready-made databases of BMCCs and energy, transportation, etc. It is important to ascertain the compatibility of these data with the goal of the study.
- Because the total life cycle includes processes that have already taken place and are known, as well as processes that will take place in future, the assumptions used in the assessment should always be clear. Preferably, separate life-cycle stages are shown.

- Because the energy consumption during use is often pronounced in the final results, the way energy consumption is assessed should be clear.
- The situation of design of a new construction using prospective LCA data (prognosis with simplifying assumptions) should be clearly distinguished from refurbishment, where the starting point is the diagnosis of the existing object.

What is typically included in the model life cycle?

International Energy Agency Annex 31 (IEA 2001) provides a description of how several LCA building tools model the life cycle of a building. These results are combined below with findings from other studies (Scholten et al. 1999; TemaNord 1995; CIB 1997; Häkkinen 1994; Mak et al. 1997; Olive 1999) as well as a Building Research Establishment (BRE) survey carried out within the Society of Environmental Toxicology and Chemistry (SETAC) Working Group on LCA in Building and Construction, which focussed specifically on the end-of-life waste management phase. This showed that there are differences in the inclusion and exclusion of life-cycle stages. The following is a summary of what has been found:

- The production phase of materials and products is included in almost all B/C LCA

Life-Cycle Assessment in Building and Construction. Shpresa Kotaji et al., editors.
©2003 Society of Environmental Toxicology and Chemistry (SETAC). ISBN 1-880661-59-7

methods and tools. In some methods, the production of components is included; in others, it is not. It is unclear which methods take all components into account and which do not.

- The transport to site of materials is included in almost all tools and methods. Transport to site of equipment and personnel is almost always excluded.
- The construction phase at the building site is not always included and, if included, often not complete for all materials due to lack of data. Construction waste is therefore also not always included. The waste of materials is sometimes included in the sense of increased specification of materials and products.
- In almost all tools the use of the building is included as energy use. Sometimes water consumption is also included.
- Maintenance and replacements get some attention in most tools. The level of detail differs.
- Demolition of the building is sometimes included. Most practitioners calculate waste arising at demolition from the functional unit or, if the scope requires, from a tonne of material. The quantity of material input is often used as an indication of waste material arising.
- Transport of waste to treatment site is almost always included. Sometimes interventions and operating impacts of waste treatment are included, sometimes not or partly. Data availability is often the bottleneck. Detailed, material-related waste treatment data typically are not available.

Several case studies reported by the SETAC workgroup members are listed in Appendix D. Some general conclusions from the case studies include the following:

- In building, the energy use during the use phase is dominant in comparison to energy-related environmental effects of building materials and products on most of the environ-

mental impact categories. The percentages differ per country and are sensitive to the life-cycle inventory (LCI) data source (Lalive d'Epinay 2000). Typical estimations give the material contribution from 10% to 20% (conventional), up to 40% or more (low-energy consumption buildings).

- Emissions to land and water and waste are often more material related. The embodied material contribution is much higher to these impact categories.
- The overall sensitivity to a change of database (LCI data source) is usually very large.
- An important conclusion of the IEA work is that the models for service life and end-of-life scenarios are the most important issues in the description of a building's life cycle. We explain this further below.

Service life

There are 4 main issues in service life:
1) the designed service life of a B/C,
2) the service life of the constituent components (BMCC),
3) replacements, and
4) the nature and frequency of maintenance.

1: Designed service life of a B/C

Because the maintenance and replacements are dependent on the 'designed' service life of the B/C, it is important to be clear about the service life chosen. There are no general rules. In fact, the designed and actual service lives are not only country dependent, but also regional or local dependent. Furthermore, the designed as well as the actual service life depends on the type of building: whether it is a monument, a flexible building, a fashion-dependent building, etc. In the Netherlands, for example, a 'standard' design service life of 75 years is chosen as a default for dwellings and 20 years for offices (this service life can be adapted by designers) (Mak et al. 1997; Stichting NVTB Projecten 2000). In the UK, 60 years is chosen as a default standard for commercial and domestic building (Howard et al. 1999). These default design service lives are derived from experience of actual service lives, although these may differ in practice, depending on the maintenance level, etc. Most LCA software tools (e.g., Envest and Eco-Quantum) also allow adaptation by the designer. In the French tool, EQUER, duration is requested from the user for the analysis instead of providing a default service life. In Finland, the LCA HOUSE tool uses 100 years as a default value (see: Appendix D). The tool also includes default values for the renewal periods of products used in different building parts, which can be changed by the user. This is also the case for the Swiss LCA tool OGIP, which uses a default value of 80 years (R.I.C.S. 1992). In Envest, the default renewal periods for BMCCs are currently fixed.

In general, the actual life cycles of office buildings are shorter than those of dwellings. This is most often due to a change in user demands. No fundamental design or durability constraint dictates this outcome. Default design service lives in LCA should preferably reflect this difference and be based on realistic models.

It is recognised that further knowledge about the relation between the service life of a building and the situation and type of building is important information for LCA in B/C. The work for the new International Organization for Standardization standard ISO 15686 for Service Life Planning may help. The standard includes service life prediction procedures, data requirements, and a framework for the incorporation of LCA data and aspects of environmental impacts. A methodology is provided to forecast the service life and estimate the timing of necessary maintenance and replacement of components. The method has been applied to test the effect of various parameters (Hakkinen et al. 2000; VTT 2000).

The decision to demolish a building generally depends on nonphysical criteria (obsolescence of some kind, value of land, etc.). There should be no confusion between service life, how it is used in industry or economical lifetime, and the assumed building lifetimes. State-of-the-art software of building LCA should include both the choice of a building lifetime (with default value) and scenarios for component lifetimes and replacement.

2: Service life of products and components

The service life of a product or component in relation to the service life of a B/C determines the number of replacements. The service life of products and components is therefore a very important parameter in LCA. In Chapter 2, we stated that a producer of a product or component must provide information about the potential service life together with the relevant boundary conditions, preferably outlined according to ISO 15686-1, for example.

There is a difference between service life (see definition list) and design life: the service life intended by the designer. The service life can be determined, for example, as the technical service life, the experienced (economic) service life, etc. There is not one rule for this: The choice depends on the goal of the study and on many factors surrounding the building. However, we see a tendency to use the demonstrated economical service life for this type of product as the default because it is the best guess for the actual service life of the product, even though it depends on the future economical situation of the B/C's user. In the Netherlands, in Switzerland, and in the UK, there are lists that can be used for this purpose (Weibel and Stritz 1995; Kohler and Zimmerman 1996; Llewellyn and Edwards 1997). However, it is also recognised that this service life (durability of components) needs much more knowledge for proper use in LCA of buildings. In the UK, a Whole Life Costing Forum has been established by representatives from the construction industry and major clients, with the aim of benchmarking typical costs and service lives. In the Netherlands, it is recognised that technical service life might be appropriate for legislative purposes because new products do not have an economical service life at introduction (Scholten et al. 1999). In Finland, a computing system has been developed for the estimation of the service life of components (Hakkinen et al. 2000).

'Performance' can be defined as a qualitative level of critical properties (ISO 15686-1). The level of properties (sometimes also called 'durability') decreases over time. At a certain moment, the decrease will no longer be acceptable, and maintenance or replacement will occur. If the time-dependent decrease in performance is related to environmental performance (as may be the case for window thermal performance or boiler efficiency, e.g.), there are more consequences for the LCA than maintenance and replacements. In such situations, the LCA process tree should be modelled in a dynamic way. However, steady-state modelling is the current practice in LCAs in the building sector.

3: Replacements

The service life of a component in relation to the service life of a B/C determines the number of replacements. It often happens that a component is replaced and that the assumed service life of the replaced component then exceeds the designed life of the building (the replacement 'outlives' the building). This remaining component life can be taken into account or not. If it is, a fraction is used as the replacement component life to be attributed to the building. If the component is not prorated against the remaining building life, then the multiplier will always be a whole number. Because the number of replacements is directly connected to the number of life cycles (production, transport, construction, demolition, waste treatment) of the component, it is clear that the choice for prorating or not can cause significant differences in LCA results.

Example: If a component with a service life of 40 years is prorated in a construction with a planned service life of 60 years, the component is used (60/40) 1.5 times (0.5 replacement attrib-

uted). If not prorated, then 1 replacement = 2 times the life cycle of the component.

An argument against prorating is that it does not reflect the true activities (1 replacement is practice; 0.5 replacement is a theoretical calculation number). Transportation of materials to the building and from the building to the disposal sites is either neglected or estimated against reference scenarios.

An argument for prorating is that it reflects average situations: there are high uncertainties in the service life of the component, the building, and the exact year of replacement. Using a fraction reflects these uncertainties. In this approach, one should not account for residual values at the end of the building's life. However, a priori residual values should not be used when prorating is not used. This subject has to be dealt with in the end-of-life scenario description.

Both prorating and not prorating have pros and contras. In some industry-initiated projects such as the Dutch MRPI (Stichting NVTB Projecten 2000), prorating is not applied. However, the BRE recently changed to this approach (Howard et al. 1999). In LCA software tools for buildings, prorating is often applied. We see a tendency to dynamic modelling, which implies that prorating should be used. More knowledge about service lives and uncertainties is required to make a proper choice.

It is important that the consequences of a choice are fully understood by the LCA practitioner. Service life planning may in the long run enable planners to better determine the actual service life and also the remaining value of a component (if any).

4: Nature and frequency of maintenance

Maintenance, like replacement, is a very important factor in LCA of buildings. Maintenance causes material use but also determines the service life of a component or even a B/C.

It is current practice to include only functional maintenance in LCAs and to exclude purely aesthetic maintenance and maintenance due to incidents and improper use because the latter are totally uncertain.

The way the frequency is determined is not always clear. This can be described by the producer (in certificates, e.g.) or can be estimated based on experiences of users. Furthermore, there is a difference in the system boundaries of the maintenance processes (see next section), and data are often lacking. The same is true for whole-life costing studies. It can be concluded that, although the importance of maintenance is recognised for LCAs, there is a lack of knowledge of how to deal with it in a proper way. An example of dealing with it is given by Häkkinen (1999).

End-of-life scenarios

End-of-life scenarios are scenarios that divide waste streams into streams that are sent to landfill, to incineration (with or without energy recovery), or to recovery (e.g., x% incineration, y% recycling). The environmental effects of waste treatment processes differ widely and so do the LCA results of different scenarios. Furthermore, end-of-life scenarios are an important policy issue and marketing instrument in many European countries.

Industry-initiated projects to date have taken 2 approaches. One approach is to define scenarios

that are based on future situations (in some situations allowed in MRPI [Stichting NVTB Projecten 2000]); others tend to define actual scenarios that are already proven (e.g., BRE Environmental Profiles, Eco-Quantum and VTT Environmental profiles/RTS environmental declarations [Häkkinen 1994; Olive 1999; Howard et al. 1999]). Both viewpoints have in common that arguments and proof must be available; scenarios on 'recyclability' often are not accepted. This recognizes the fact that, for many materials, producers can claim 'it can be recycled' and that this claim differs from having the statistics and evidence of recycling taking place. It therefore can be difficult to differentiate between products that have not yet been available for long enough to enter the waste stream in economic quantities, but that technically are recyclable, and those that are available but not recycled because it is not economical to do so under current market conditions. Most LCA methods recognise the benefits of using recycled or secondary materials in products today, rather than the benefits of recycled material being available in the future.

End-of-life scenarios differ per country but may also differ per B/C (Amt für Bundesbauten 1995). This must be taken into account when an LCA is performed.

The sensitivity for end-of-life scenarios is high. A Swiss example of office buildings (Lalive d'Epinay 2000) showed that in a worst-case end-

There are not yet commonly accepted rules as to how to define B/C service life or BMCC service life, frequency of maintenance, and replacement. Development in service life planning to determine actual service life and remaining value of the BMCC will be useful.

Actual, or economic, service lifetimes, based on experience, generally are favoured over potential or maximum technical lifetimes.

Dynamic modelling of environment-related durability of BMCCs has been neglected to date and should receive more attention.

of-life scenario, the disposal phase is as relevant as the construction phase of the life cycle of the building (it is a building with a very good energy performance): construction phase 37.4%, use phase 26.1%, demolition phase 36.5% (assessed with Eco-indicator 95). This contribution is 36% higher than an optimised disposal scenario for the demolition phase of the building.

Alternative disposal scenarios are also studied by Häkkinen (1999). Other examples can be found in various case studies (see Appendix D).

Due to the lack of data available on the impacts of specific materials in disposal, the LCA practitioner has 3 choices:

- To model the impact of the disposal specifically for different BMCCs, often incompletely.
- To ascribe general impacts on a mass basis from a national model of waste disposal methods.
- To recognise 'waste produced' as an endpoint, with the removal of recycled materials from this flow.

Recycling processes must be considered for implications on the boundary of the study. See the discussion in 'Allocation' (p 27).

The BMCC–LCA Life Cycle: System Boundaries and Completeness of Inputs and Outputs

Each BMCC–LCA has its own life cycle, which may be described by a process tree. The system boundaries as well as the completeness of inputs and outputs of unit process must be consistent for all BMCC–LCAs in order to be able to add the BMCC–LCAs. Therefore, detailed system boundaries and requirements for completeness are defined in several projects. This state-of-the-art report summarises such system boundaries from different projects (Tables 4-1, 4-2, 4-3). It is too early to give guidelines, but we try to show general trends that may later be developed into guidelines.

Ending the life cycle with the generation of waste streams or further accounting for waste treatment are both frequently applied in BMCC–LCAs and B/C–LCAs. The waste scenario can be formulated in 2 ways, depending on the study goals:

- The currently applied waste scenario.

- A hypothetical waste management scenario (e.g., based on developing recycling technologies).

If the current scenario is chosen, it is useful for designers to also be able to apply future scenarios, for example, for a design-for-recycling study.

End-of-life scenarios are country dependent and sometimes building dependent. It is useful to have the freedom to choose the appropriate scenario as well as a common national default scenario.

Modelling of waste disposal impacts is currently limited by data availability, and a number of solutions to this problem are found in current state-of-the-art methods.

Table 4-1 Overview of boundaries applied in different projects in the Netherlands, the UK, and Switzerland

	Netherlands, MRPI project (Stichting NVTB Projecten 2000)	United Kingdom, BRE project (Howard et al. 1999)	Switzerland, thesis ETH (Lalive d' Epinay 2000)	Switzerland, OGIP (OGIP 1998)
INPUTS				
Resources (raw materials) of a unit process taken into account	100% of the resources must be described qualitatively, at least 95% by mass must be analysed quantitatively (production process) and the environmentally relevant production processes.	98% by mass must be analysed quantitatively.	Include resources for material production, energy production, and demolition processes (emissions of demolition). Exclude building construction processes (on site).	Resources for material production, energy production, and demolition (only amount of material) are included. Building construction processes (on site) are partly included.
Transport of resources of a unit process to plant	Type of transport, distance, and return transport based on actual situations, distinction between mass and volume transport.	Type of transport, distance, and return transport based on actual situations, distinction between mass and volume transport.	Type of transport, distance, and return transport based on actual situations, distinction between mass and volume transport. Transportation of material to the building and disposal sites is or neglected or estimated with a reference value.	Type of transport, distance, and return transport based on actual situations, distinction between mass and volume transport. The transportation of the building material to the build-ing and from the build-ing to the disposal site is estimated with a reference value.
Energy inputs of a production process	All, except for space/office heating. Energy input is a resource: production is accounted for as secondary energy (definition by IFIAS). Calorific value of secondary energy carriers included. Lower/higher heating values not defined. Feedstock energy of raw materials is excluded.	All, including space/office heating. Energy input is a resource: production is accounted for as primary energy (definition by IFIAS). Calorific value of secondary energy carriers included. Lower/higher heating values defined. Feedstock energy is included for fossil fuels, excluded for non-economic fuels.	All, including space/office heating. Use data and methodology of Frischknecht et al. (1996) to include energy in/out. Uses data and methodology of Weibel and Stritz (1995) and Doka (2000) to include production and disposal.	All, including space/office heating. Uses data and methodology of Frischknecht et al. (1996) to include energy input. Uses data and methodology of Weibel and Stritz (1995) to include material production.
Water consumed per unit of production	Includes only energy used for extraction.	Included.	Includes only energy use for purchasing.	Only energy use for purchasing is included.
Capital equipment (production, maintenance, disposal)	Excluded, except for frequently consumed items	Excluded, except for frequently consumed items.	Not mentioned.	Not mentioned.
OUTPUTS				
Emissions to air/water per unit process	Process-specific emissions according to environmental permits and listed energy emissions Completeness: 95% of all emissions by mass and environmentally relevant emissions	Emissions as measured by manufacturers and fuel emissions.	Emissions to ICP or other lists.	Emissions to ICP or other lists.
Emissions to land/waste per unit process	Waste treatment must be in-cluded as a process. Inputs and outputs of waste treatment pro-cess (as stated above). Only final (landfilled) waste is waste. Type (dangerous or not, according to national law) must be defined. Completeness: 95% by mass of all waste streams must be included.	Waste produced is measured and recycled materials re-moved. Emissions of landfill and incineration are restricted to CO_2 and CH_4 generated by timber (further work follows).	Emissions to land. Waste processes are included in form of emissions from landfilling and/or burning of building products.	Emissions to land. Waste produced is measured and recycled materials removed.
VALIDATION CHECK				
Mass balance per unit process	Yes	Yes	Yes	Yes
Energy balance per unit process	No	No	Yes	Yes
Mass balance: company level	Yes	Yes	Not mentioned	Not mentioned
Energy balance: company level	Yes	No	Not mentioned	Not mentioned
External	Yes	Yes, for external claims	Yes	Yes

Table 4-2 Overview of boundaries applied in different projects in Finland, Canada, Denmark, and France

	Finland, VTT projects (Häkkinen 1994)	Canada, ATHENA Project (SBR 1998)	Denmark (Holleris 2000; Holleris et al. 2001)	France, EQUER project (Peuportier 1998)
INPUTS				
The resources (raw materials) of a unit process that are taken into account	All the resources must be described qualitatively and, in principle, also quantitatively. If some minor parts of material flows are excluded, the significance is reasoned.	Resources for material production, energy production, and building construction process (on site) are included. At least 95% by mass and energy is accounted for, as are any environmental relevant production processes	All resources must be described qualitatively and, in principle, also quantitatively.	Materials and components present in small quantities (e.g., auxiliary materials) may be neglected, in practice, around 95% by mass is analysed (responsibility of the user).
Transport of resources of a unit process to plant	Type of transportation distance based on actual situations. Consumption of fuels per tonkm on the basis of typical values for the types of products in question. Environmental profiles of fuels based on national average.	Type of transport, distance, and return transport based on actual situations and/or industry averages. The transportation of the building material to the building site is also included, as is transport of heavy equipment to and from the construction site.	Type of transportation and distance are based on actual situations from raw materials extractions to gate. From gate to construction site, average distances are estimated.	Default transport distance and type is proposed to the users (e.g., 100 km truck between fabrication plant and building site).
Energy inputs of a production process	All, including the heating of space. All the stages from raw material extraction onwards are included with regard to fuels as well as electricity and district heat. Feedstock energy is included. Energy is classified in renewable and nonrenewable resources.	All, except for space/office heating. Primary energy input included as a resource. Feedstock energy is included for fossil fuels, excluded for non-economic biomass based fuels.	All, including the heating of space. All the stages from raw material extraction onwards are included with regard to fuels as well as electricity and district heat. Energy is classified in renewable and nonrenewable resources.	All, including space/office heating (link with multizone simulation), and if needed (according to the goal of the LCA study), electricity consumption, heat for domestic hot water and cooking, heat recovery from domestic waste. Energy is considered in 2 indicators: exhaust of natural resources and primary energy consumption (upper heating value, feedstock energy considered).
Water consumed per unit production	Normally excluded.	Included.	Normally excluded.	Includes (according to the goal) possibly the water consumption during use.
Capital equipment (production, maintenance, disposal)	Excluded.	Excluded.	Excluded.	Excluded, except for energy and transport processes.
OUTPUTS				
Emissions to air/water per unit process	Both process and fuel-based emissions are included. When producing data for environmental declarations, the emissions are classified and characterised.	Detailed fuel and process emissions according to ATHENA LCI methodology. Provide characterisation summary measures at the building level.	Both process and fuel-based emissions are included.	At the product level only. CML indicators are considered at the building level.
Emissions to land/waste per unit process		Waste treatment is accounted for at the unit process level. Only final (landfilled) waste is waste.	Only final (landfilled) waste is waste. Waste and recycling materials are calculated.	Total waste mass, including fabrication, construction, use, renovation, and demolition concerning building materials, and total building mass are considered. Waste treatment processes are included in the model but not in the applications because of lack of available data.
VALIDATION CHECK				
Mass balance per unit process	Yes	Yes	Yes	In one case study for the whole building, not for BMCC–LCAs.
Energy balance per unit process	No	Yes	No	In one case study for the whole building, not for BMCC–LCAs.
Mass balance: company level	Yes	Yes, but not on a consistent basis.	Yes	No
Energy balance: company level	?	No	No	No
External	No	Yes	No	No

Table 4-3 Overview of boundaries applied in different projects in Germany and Norway

	Germany, GaBi Project (Stuttgart University) www.gabi-software.com	Norway, NBI project (Petersen and Fossdal 1998)
INPUTS		
Resources (raw materials) of a unit process that are taken into account	Min. 99% of input flows (energy and mass) are included at every unit process back to the resources. Energy resources are country specific (e.g., resource natural gas from the Netherlands).	Raw material excavation, production, building site, use and demolition processes are (or can be) included.
Transport of resources of a unit process to plant	All transport is included (cradle-to-gate), taking into account type of transport, average distance in Germany for specific step, average utilisation of the means of transport, way back.	All transportation is covered, based on key figures and travel distance (tonkm).
Energy inputs of a production process	All energy related to the production of the product is included (e.g., machinery, heating of the factory floor or the production office); administration and sales department is excluded.	All included; electricity consumption is based on national grid. Lower/higher heating values not defined. Feedstock energy of materials is included.
Water consumed per unit of production	Included.	Included.
Capital equipment (production, maintenance, disposal)	Excluded.	Excluded.
OUTPUTS		
Emissions to air/water per unit process	Detailed emissions to air, water, and soil are included in the calculations; the emissions are classified and characterised.	Included.
Emissions to land/waste per unit process	Waste is reported in 4 classes: spoile pile, ore processing residues, consumer waste, and hazardous waste. Directly recycling in the production process is included.	Included.
VALIDATION		
Check of mass balance per unit process	Yes	No
Check of energy balance per unit process	Yes	No
Check of mass balance on company level	Yes	No
Check of energy balance on company level	Yes	Yes
External validation	No	?

LCI Data

Appendix D provides an overview that is the result of a literature survey carried out by the workgroup. Appendix D lists a wide range of LCI and life-cycle impact assessment (LCIA) studies, which have been carried out by different organisations for B/Cs and BMCCs.

Allocation

Allocation (the partitioning of environmental impacts between systems) is a general LCA issue and discussion point in the whole LCA arena.

Because of the long life of B/Cs, allocation for end-of-life recycling needs special attention in the B/C sector. Allocation is necessary for multi-output systems, for multi-input systems, and for waste treatments. Methodological details can be found in ISO 14041 and the ISO Technical Reports.

Extremes for end-of-life recycling

For end-of-life (recycling) allocations, allocation methods vary from a cutoff of the recycling processes (recycling not taken into account) to a full subtraction of avoided impacts (which is the same principle as expanding system boundaries, suggested by ISO 14041 as one of the possibilities to avoid allocation). Most other methods (e.g., economic-cut, value-corrected substitution) result in scores that fall between these extremes.

In the cutoff method discriminates, the possible benefits from recycling are not taken into account. The recycling processes are fully allocated to the next product system (the system where the secondary material is used). This allocation is then reflected in the environmental impacts attributed to the use of secondary materials. Such an approach clearly has implications for the use of LCA as a policy tool to promote the use of recycled or secondary materials.

A full subtraction means that the benefits of (assumed) future recycling are already regarded as benefits for the present situation. The current environmental impacts are allocated to a future generation. Subtraction usually does not take technological changes into account. The present production processes are subtracted instead of the future, unknown but possibly cleaner, technologies. For long-living products such as B/Cs, it is crucial to think about the way to deal with future recycling scenarios because these are often uncertain.

A choice between allocation methods depends on the goal of the study. For example, allocation methods for policy LCAs may differ from those for a comparative LCA. Therefore non-allocated data are preferred in databases to allow decision-makers to make their own allocation choice. In practice, this may not always be possible.

Definition of end-of-life scenario

The last part of 'The Building Life Cycle' (p 19) deals with the definition of end-of-life scenarios. A clear definition is necessary in combination with allocation methods for BMCC– and B/C–LCAs. This is particularly true for quantities of secondary material that are used today and the material to be recycled in the future.

Many tools and models in the B/C sector are based on actual (today's) situation concerning the percentage of recycling. Some tools also use actual percentages of secondary material as material input (Mak et al. 1997; Howard et al. 1999); others also allow guaranteed percentages for the future (Stichting NVTB Projecten 2000). Discussions about how to deal with growing recycling markets (which cause an imbalance between recycling percentage and secondary input percentage) are still going on.

Allocation in BMCC–LCAs for B/C

We have decided to focus on the problems that are encountered when adding BMCC–LCAs to B/C–LCA. The main point of attention, as mentioned in several studies (Mak et al. 1997; Howard et al. 1999; Stichting NVTB Projecten 2000), is the so-called '100% rule' of ISO. The 100% rule implies that the sum of the allocated systems equals the unallocated system. When carrying out a BMCC–LCA, the 100% rule (ISO 14041) must not only be guaranteed on the level of the BMCC–LCA but also must fit into a 100% rule on the B/C level. Problems are identified for allocation between BMCC–LCAs and for addition of BMCC–LCAs that apply other allocation principles.

If the system boundaries of a BMCC–LCA are expanded to avoid allocation, the system boundaries of this BMCC–LCA may not comply with the system boundaries of other BMCC–LCAs. This may hamper the combination of BMCC–LCAs for B/C–LCAs. Such LCAs may, however, be useful for policy decisions.

Example: An example of system expansion is to include a waste incinerator of household waste within the boundary of a BMCC–LCA that uses secondary fuels that can also be incinerated in that waste incinerator. Another example, again from the slag cement industry, but different to that above, is the inclusion of slag cement production in the steel-BMCC–LCA in order to avoid allocation between steel and cement. The environmental impacts of the avoided products in cement production are in this case subtracted (or value-based subtracted) from the steel-BMCC–LCA.

If one BMCC–LCA applies a cutoff allocation method for the recycling process, while another BMCC–LCA applies a subtraction method, it can be questioned whether the system boundaries are equal and thus, whether these BMCC–LCAs can be combined.

Example: When stony materials apply economic cutoff and metals apply (value-corrected) substitution by regarding the system as a closed loop.

Current allocation discussions in the B/C sector mainly deal with the applicability of system expansion and, as a consequence, with the comparability of cutoff methods versus subtraction methods.

Most independent instruments and tools used for different material types in the B/C sector, especially on the B/C level, tend to favour exclusion of system expansion and of subtraction because of the incomparable system boundaries. However, some LCA practitioners, mostly outside of the B/C industry, still claim that the addition of LCAs with expanded system boundaries is possible.

System boundaries and allocation rules differ between projects and countries. Although there is an international knowledge transfer and growing consensus, further harmonisation is desired for information transfer of LCA data between studies and countries.

End-of-life allocation

Different views exist on how to deal with allocation for long-life applications such as building products and B/C. Main discussion points include:
- The applicability of system expansion to avoid allocation and to apply subtraction methods, especially with regard to the comparability of BMCC–LCAs for combination in B/C-Class.
- The applicability of subtraction methods for (semi-)closed-loop recycling processes, also especially for BMCC–LCAs to be combined in B/C-Class.
- The percentage of recycled material to allocate to the current product that is being produced. There are 2 main views:
    1) Take the impact based on the current percentage of recycled material only.
    2) Treat virgin product today as recycled because it is going to be recycled.
- The fact that allocation procedures concentrating on the devaluation ('quality loss') of a material over a life cycle can be more appropriate than using factors on recycling rates.

Although this discussion is not building specific (it is relevant for all long-life products), it is essential for LCAs of B/C because the use of BMCC–LCAs in B/C–LCAs for comparative purposes is possible only when comparable allocation principles are applied.

Multi-input and multi-output processes

It is the responsibility of the LCA practitioner to use the appropriate allocation principle for products. The allocation method should reflect the causality of material and energy flows.

Allocation principles in strategy and policy

LCAs may differ from those in B/C LCAs. Ideally, LCA data on end-of-life processes should be provided in a non-allocated format to allow decision-makers to choose the appropriate allocation method for the goal of the study.

On the other hand, subtraction methods are identified as useful in LCAs for strategy and policy discussions. The comparability of the system boundaries is still the subject of international discussion.

Multi-input and multi-output processes

Although not a specific B/C subject, it is worthwhile to pay a little attention to this subject. We found different allocation principles for multi-input and multi-output processes in BMCC and B/C tools and instruments. Multi-output processes may be allocated on a mass basis as a default, but there are also methods using economic-based allocation. Sometimes physical or chemical allocation bases are applied. Multi-input processes often tend to physical and chemical allocation bases.

Example: Blast furnace slag from steel production is applied in cement. If, for example, the cement industry considers blast furnace slag a waste material and includes it for free in the cement-BMCC–LCA, the production of slag should be totally allocated to the production of steel. However, if the steel industry argues that slag is not a waste, and allocates environmental impacts to the slag, the sum of the cement-BMCC–LCA and the steel-BMCC–LCA does not equal the sum of the unallocated steel and cement production processes. Both BMCC–LCAs cannot be combined in a B/C–LCA.

There is not 'one choice' in this. ISO also provides several possibilities. However, allocation principles in BMCC–LCAs for combination in B/C–LCAs must be as consistent as possible, and further progress towards this goal is desirable. If there are valid arguments not to use an equal allocation basis for all BMCC–LCAs, deviations in allocation basis can be possible.

References

Amt für Bundesbauten. 1995. Nutzungszeiten von Gebäuden und Bauteilen. AfB, Bern, CH: Amt für Bundesbauten.

[CIB] International Council for Research and Innovation in Building and Construction. 1997. Buildings and the environment. Second International Conference. Paris, F: CIB.

Doka G. 2000. Ökobilanzen der Entsorgung von Baumaterialien. Zürich, CH: ETH Zürich, Gruppe Für Sicherhei und Umweltschutz ZEN, Empa Dübendorf.

Frischknecht R, Bollens U, Bosshart S, Ciot M, Ciseri L, Doka G, Dones R et al. 1996. Ökoinventare von Energiesystemen. Grundlagen für den ökologischen Vergleich von Energiesystemen und den Einbezug von Energiesystemen in Ökobilanzen für die Schweiz. 3. Auflage, Gruppe Energie - Stoffe - Umwelt (ESU), Eidgenössische Technische Hochschule Zürich und Sektion Ganzheitliche Systemanalysen, Paul Scherrer Institut Villingen/Würenlingen, CH: ETH Zürich.

Häkkinen T. 1994. Environmental impact of building materials. Espoo, SF:VTT. Research notes 1590.

Häkkinen T. 1999. Environmental impact of coated exterior wooden cladding with specific reference to service life. Espoo, SF: VTT. http://www.vtt.fi/rte/projects/environ/enviro_prj_paints.html. Accessed 23 Mar 2003.

Häkkinen T, Vares S, Vesikari E, Karhu V. 2000. Product information for service life design. Espoo, SF: VTT. http:/www.vtt.fi/rte/projects/environ/enviro_prj_tuki.html. Accessed 23 Mar 2003.

Holleris Petersen E. 2000. Building environmental assessment tool-BEAT 2000. Proceedings International Conference Sustainable Buildings 2000; 2000 Oct 22–25; Maastricht, NL.

Holleris Petersen E, Dinesen J, Krogh H. 2001. Environmental data for building elements (in Danish with an English summary). Horlsholm, DK: Danish Building and Urban Research.

Howard N, Edwards S, Anderson J. 1999. BRE methodology for environmental profiles of construction materials, components and buildings. London, GB: Building Research Establishment (BRE).

[IEA] International Energy Agency. 2001. Existing instruments, tools and databases. Annex 31. Available at www.uni-weimar.de/SCC/PRO/ and http://annex31doc.tce.rmit.edu.au/. Accessed 6 December 2002.

Kohler N, Zimmerman M. 1996. OGIP/DATO Optimierung von Gesamtenergieverbrauch, Umweltbelastung und Baukosten. Karlsruhe, D: Universität Karlsruhe, Institut für Industrielle Bauproduktion (IFIB).

Lalive d'Epinay A. 2000. Die Umweltverträglichkeit als eine Determinante des architektonischen Entwurfs. ETH Zürich: Abteilung für Umweltnaturwissenschaften,

Laboratorium für technische Chemie, Gruppe für Sicherheit und Umweltschutz. Diss ETH Nr. 13610.

Llewellyn J, Edwards S. 1997. Towards a framework for environmental assessment of Building Materials and Components. London, GB: CRC Ltd. ISBN 1 86081 252 X. (CRC@Construct.emap.co.uk)

Mak J, Anink D, Kortman J, van Ewijk H. 1997. Eco-Quantum computer programme. Gouda/Amsterdam, NL: W/E Consultants Sustainable Buildings/IVAM. www.ecoquantum.nl. Accessed 23 Mar 2003.

[OGIP] Optimierung von Gesamtenergieverbrauch. 1998. Umweltbelastung und Baukosten, Universität Karlsruhe, Institut für Industrielle Bauproduktion (IFIB), Karlsruhe, Germany: Universität Karlsruhe, Institut für Industrielle Bauproduktion (IFIB), OGIP/DATO. www.ogip.ch.

Olive G. 1999. Ateque: 5 ans de travaux (1993–1998). Paris, F: PUCA Edition ref. 109.

Petersen TD, Fossdal S. 1998. Environmental declaration of building materials, constructions and technical installations, a part of NBI Technical Approval. Oslo, N: NBI.

Peuportier B. 1998. The life cycle simulation method EQUER applied to building components. CIB Conference Construction and the environment; 1998 Jun; Gävle, S.

[R.I.C.S.] Royal Institute of Chartered Surveyors. 1992. Life cycle costing in buildings: Keynote lecture notes. Brighton, GB: RIC.

SBR. 1998. Service life of building products. Values from practice. Rotterdam, NL: SBR. ISBN 9053672591.

Scholten NPM, De Groot-Van Dam A. 1999. Material-related environmental profile of a building. A prototype method. Delft, NL: TNO Bouw, a joint cooperation with IVAM Environmental Research, W/E Consultants Duuzaam Bouwen, TNO MEP, CML, INTRON. TNO rapport 1999-BKR-R025.

Stichting NVTB Projecten. 2000. Manual and background document for compiling environmentally relevant product information (MRPI), SNP-R98002 version 1.2. Driebergen, NL: Stichting MRPI.

TemaNord. 1995. Environmental data for building materials in the Nordic countries. Copenhagen, DK: TemaNord 1995:7.

VTT. 2000. Requirements management tool-EcoProP Software. Espoo, SF: VTT. Available at http://cic.vtt.fi/eco/ecoprop/. Accessed 23 Mar 2003.

Weibel T, Stritz A. 1995. Ökoinventare und Wirkungsbilanzen von Baumaterialien. Zurich, CH: Eidgenössische Technischen Hochschule Zürich (ETHZ) und dem Bundesamt für Energiewirtschaft (BEW).

Weibel T, Stritz A. 1995. Ökoinventare von Baumaterialien, Institut für Energietechnik, Laboratorium für Energiesysteme, Gruppe Energie-Stoffe-Umwelt. Zürich, CH: ETH Zürich.

5 Impact Assessment

The impact assessment step in life-cycle assessment (LCA) for building and construction (B/C) is the same as for other LCAs. It is the step in which quantitative results of the inventory analysis are evaluated and aggregated into environmental loads.

Life-cycle assessment usually focuses on external regional and global environmental effects without considering how these effects are distributed in time and space. However, there are important environmental problems related to B/C that arise locally, such as the emissions of dangerous substances into the indoor climate, which can affect human health or the impacts of land use and disturbance to ecosystems.

Local Emissions

During the B/C stage, several types of substances can be emitted and can affect workers' or building occupants' health. For example, B/C workers can be exposed to dangerous organic or inorganic substances released during the handling of material or products, or they could be exposed to wood sawdust, for example. Certain building products and materials may contain substances that are added to affect their properties (fire retardants, plasticizers, biocides, etc.). As shown in

Table 5-1, those substances can be classified as hazardous and can be emitted in certain circumstances (Krogh 1999). They can impact the working environment during construction works or the indoor air during building use, or they can be released during waste disposal.

Life-cycle assessments usually focus on external regional and global environmental effects without considering how these effects are distributed in time and space. Impacts on human health from the locally released chemicals are difficult to assess for 2 key reasons:

1) Data are missing
   - for the production of the chemicals (databases give only the consumption of energy and some aggregated emission data),
   - for emissions to the working environment,
   - for emissions to indoor climate (some data exist from indoor labelling), or
   - for disposal processes (a few leaching data exist).
2) The level of emission can be such that typical LCA results, from cradle-to-grave, will not be capable of highlighting potential concerns for human health during the use or construction phase.

Table 5-1 Examples of harmful substances in building products, recognised as causing effects on human and environmental health

Categories	Substances	Building products	Problems
Metals	Arsenic compounds Lead and lead compounds Cadmium Chromium compounds Organic tin compounds Nickel Copper compounds	Impregnated wood Coverings, wiring, polyvinyl chloride (PVC) Pigments, soldering Impregnated wood Impregnated wood Locking device Impregnated wood	Metals are not biodegradable. Use of products containing metals can cause emissions to the environment that accumulate there and at last appear in the food chain, with risk for human health effects.
Not easily degradable compounds	Polychlorinated biphenyls (PCBs) Phthalates	Sealants Sealants and plastic products	Substances will accumulate in the environment and at last appear in the food chain, with risk for human health effects.
Solvents		Paints, impregnating oil	Solvents emit to the working environment and can cause health effects for workers.
Dispersing agent	Nonylphenolethoxylates	Paints	Emitted to the aquatic environment, where it can have health effects for water organisms.
Biocide	Fungicides Conservation agents	Sealants, paints Sealants, paints	Emitted to the aquatic environment, where it can have health effects for water organisms.
Monomers	Isocyanate Epoxy compounds Phenol Formaldehyde	Sealants Epoxy adhesives Two-component adhesives Two-component adhesives	Monomers react and form polymers. The compounds can emit to the working environment and cause human health effects for workers.
Others	Borax Boric acid	Insulation materials	Borax and boric acid may cause sterility and damage to the capacity for reproduction.
Wood	Sawdust	Beams, panels	Causes human health effects, especially from hard wood dust.
Mineral fibers	Asbestos	Secondary aggregates	Though illegal as a building material in many European countries, they often are found in secondary aggregates.

Indoor air quality

'Indoor climate' is a term used to describe physical properties of the indoor environment (temperature, humidity, lighting, noise levels, concentration of particles or gases, etc.) which may affect the health and/or comfort of the building occupant. Indoor air quality (IAQ) is one property of the indoor climate. IAQ is affected by concentration of dust and particles, aerosols, inorganic gases, and volatile organic compounds (VOCs).

The impacts of materials on the IAQ may be crucial to the outcome of an LCA. A model is proposed for calculating the health-specific IAQ impacts of materials during their performance time. The proposed model is based on the indoor climate labelling in Denmark and Norway (formerly the Danish Indoor Climate Labelling [Wolkoff and Nielsen 1994; Jensen and Wolkoff 1996] and www.dsic.org) and using the principles of the Danish LCA model EDIP (Wenzel et al. 1997; Hauschild and Wenzel 1998). This indoor climate labelling is based on the time that concentrations of emitted VOCs need to decay below 'indoor-relevant' levels. The labelling system combines both chemical and sensory evaluation of emissions, and has been in use since 1995.

The possibility to include indoor climate issues as an impact category in LCA was investigated by Heijungs (1992). It was found that only very limited aspects could be addressed in LCA, and those issues preferably are dealt with separately, using dedicated tools such as indoor air assessment (Johnson 1998).

Working environment

The relation between LCA and the working environment is being addressed by the Society of Environmental Toxicology and Chemistry (SETAC) Working Group on LCA and the Working Environment.

Disposal

There is little information regarding the impact of substance release during disposal. Because a rise in recycled material consumption is expected, the issue of substance release could become more prominent.

There is a clear need, expressed by the various building actors, to acquire a better understanding of potential issues related to use and handling of some materials and their potential releases of dangerous substances.

At the European level, the Construction Product Directive essential requirement relating to hygiene, humans, and the environment will focus on building-use phases impacts to indoor air and outdoor environments as well as to water supply.

At national levels, for example, Norway, Denmark, and Sweden have developed lists of chemicals to avoid or to be restricted if being used (Krogh 1999; Danish Environmental Agency 2000a, 2000b; ENTREP 2000; KEMI 2000; SFT 2000).

Some countries currently include in their environmental declaration schemes (whether Type II or III) a list of chemical content and/or emissions. There is still controversy as to whether these data can be incorporated into an LCA.

It should be noted that emissions of toxic substances are not the only health issue in the B/C industry. Other issues include noise and heavy-good lifting.

Land Use Effects and Ecosystems

Land use in LCA

The use of land surface for economic purposes (buildings and traffic systems, mineral extraction, agriculture, silviculture, etc.) is probably the most important cause of the ongoing degradation of terrestrial ecosystems. Nevertheless, the occupation of land surface has not been reflected in most LCA applications to date.

Hazardous substances are not easily dealt with in B/C–LCAs or building material and component combination (BMCC)–LCAs. Under certain conditions, hazardous substances can be released from materials or formed when some materials are combined; these substances impact building IAQ or the working environment and result in health issues. IAQ is highly linked to building design that affects the pollutant dynamics.

Further discussion and research is needed to establish an appropriate treatment (if any) of IAQ in LCA.

The relation between LCA and the working environment is being addressed by the SETAC Working Group on LCA and the Working Environment.

This shortcoming in the LCA method may be more acceptable for users of LCA on products that are manufactured in complex processes, like metals and plastics, because toxic emissions and energy use tend to be regarded as the key issues here. For building materials, which are used in high volumes but need relatively little processing, the use of land surface tends to be one of the key issues in environmental decision-making. This is underlined by the policies in European countries to promote the use of secondary raw materials and wood from sustainable managed forests and the proposals for levies on mineral extraction.

Even more than land use in the production chain, local land use for a B/C project will often be relevant for decision-making. The selection of a site can have significant impact (e.g., additional infrastructure requirements). However, this will generally not be included in an LCA of a B/C. It can be assessed by a more detailed method, which takes into account the issues that are relevant in the area. (See also Chapter 8 on environmental impact assessment.) This section focuses on methods that can be applied to land use in the production chain.

The SETAC WIA-2 Task Force on Resources and Land treats land use in LCA in detail (a draft version of their final report has been a source for this section). Because of the significance of LCA in B/C, it seems to be appropriate to pay some attention to it in this report.

Researchers have published several proposals for including land use in LCA (Damen Consultants 1998). Some of them have been used in case studies, but none of the methods is in daily use today. This also means that, for example, LCA databases, which are the main source of data for LCA practitioners, usually lack land-use data. Projects to collect LCA data from industry do not include a field on land use. Land use, though a key issue in decision-making, is still in its infancy. A state of the art in LCA practice does not, therefore, really exist, but we can give some details of the state of the art in LCA research. This will be sub-

ject to change, and it is recommended that readers visit the SETAC website (www.setac.org) to follow the latest developments in this subject.

Two types of land use are distinguished. For some processes, a unit of product can be harvested from a certain area in a certain period (e.g., agriculture, silviculture). This intervention is called 'land occupation'. Some people do not regard this to be an environmental issue, but others do because this form of occupation precludes renaturalisation of the area. On a broader scale, the growing world population still occupies more land every year, which leads to less room for nature.

For other processes, an area is used for a certain period, after which the area often has a chance to regenerate (e.g., mineral extraction, waste landfill). This intervention is often called 'land transformation'. Because of changes in the physical conditions, the potential quality of the area may have been affected.

The amount of land use as such is an indicator for competition over the natural resource 'land area', which can be included as an LCA effect score. To determine the effects of land use on ecosystems, the amount and the intensity of use must be evaluated. Therefore, a quality indicator is introduced, leading to the following expressions:

Land occupation = Quality × Area × Time

Land transformation = Quality × Area

This quality indicator is one of the most difficult aspects of land use in LCA. Experts in the field of judging environmental impacts of projects tend to use a large number of functional indicators, describing effects on biotic and abiotic nature. Such an approach, though valuable and perhaps consistent with the detailed way in which emissions are treated in LCA, is hardly feasible for the everyday LCA study on a complex product because of the data needs and the need for local expertise.

Therefore, researchers developed simplified indicators for the land use. Some authors—including those of the well-known CML guideline (Lindeijer 2000)—simplified the problem by defining land-use classes (e.g., nature, extensive use up to built-up area). The major disadvantage is the neglect or exaggeration of small interventions.

Other researchers therefore sought quantitative indicators for the quality. For the quality of an ecosystem on a patch of land, one can define innumerable indicators. Most of them, however, cannot be specified without field research. Because this is hardly practical in LCA studies, it has been found necessary to choose very simple indicators that can be operationalised within a reasonable time frame. Research is focused on indicators for 2 endpoints: biodiversity and life support functions. For biodiversity, the vascular plants species density in an area is one of the most promising quality indicators. For life support, several indicators have been proposed, but we cannot yet state which of these is the most promising.

Land use also has a direct influence on man because people feel comfortable in a well-known environment, and they value the landscape (aesthetics, cultural heritage, and naturalness). Therefore, land transformation and changes in type of occupation often meet local resistance. Researchers have not yet succeeded in finding quality indicators for this influence on human welfare.

References

Danish Environmental Agency. 2000a. Statutory order nr. 733. List of dangerous substances. Available at www.mst.dk. Accessed 25 Mar 2003.

Danish Environmental Agency. 2000b. List of unwanted chemicals and Effects. Available at www.mst.dk. Accessed 25 Mar 2003.

[KEMI] Kemikalieinspektionen, 2000. OSB-liste. Stockholm. KEMI. Available at www.kemi.se. Accessed 25 Mar 2003,

[ENTREP] Entreprenørernes BST ApS. 2000. Database for safety sheet for chemicals products. Herlev, DK: ENTREP. Available at www.entrep-bst.dk. Accessed 25 Mar 2003.

Damen Consultants. 1998. Service life of building products. Values from practice. Levensduur van Bouwproducten. Rotterdam, NL: Stichting Bouwresearch (SBR).

Hauschild M, Wenzel H. 1998. Environmental assessment of products. Vol. 2, Scientific background. London, UK: Chapman & Hall.

Heijungs R, editor. 1992. Environmental life cycle assessment of products: Guide and backgrounds. Utrecht, NL: Novem.

Jensen B, Wolkoff P. 1996. VOC-database. Odor thresholds, mucous membrane irritation thresholds, physico-chemical parameters of volatile organic compounds. Version 2. Copenhagen, DK: National Institute of Occupational Health.

Johnson A. 1998. LCA and indoor climate. Technical Environmental Planning. Göteborg, S: Chalmers University of Technology.

Krogh H. 1999. Harmful substances in building products (in Danish). SBI Bulletin 113. Copenhagen, DK: Danish Building Research Institute.

Lindeijer E. 2000. Review of land use impact methodologies. *J Cleaner Production* 8:273.

[SFT] Statens Forunensningstilsyn. 2000. Miljoevernmyndighetenes OBS-liste. Oslo, N: SFT. Available at www.sft.no/publikasjoner. Accessed 25 Mar 2003.

Wenzel H, Hauschild M, Alting L. 1997. Environmental assessment of products. Volume 1, Methodology tools and case studies in product development. London, UK: Chapman & Hall.

Wolkoff P, Nielsen PA. 1994. Indoor climate labelling of building materials, chemical emission testing, modeling and indoor odor thresholds. Copenhagen, DK: National Institute of Occupational Health and Danish Building Research Institute. Available at www.dsic.org. Accessed 25 Mar 2003.

6 Interpretation

The interpretation step is not specific for the building and construction (B/C) sector. Therefore, this chapter is restricted to the identification of B/C-specific issues that are recommended for inclusion in a sensitivity analysis and a checklist to use when comparing 2 life-cycle assessments (LCAs) for building material and component combinations (BMCCs). The checklist consists of these elements:

- The functional equivalence of B/C and/or BMCCs when comparisons are made, especially from the viewpoint of the total B/C.
- The performance characteristics of the building in the functional unit.
- The life expectancy of the B/C and of the components.
- The transportation (if relevant) to the site. Transport to site is often relevant for bulk building materials (following, e.g., from CIA 1998 and Miller 1999; see also Appendixes B and C).
- The representativeness (if relevant) of data used for local materials (see Appendix C).
- The durability of components in relation to environmental performance.
- The end-of-life scenarios.
- The scenarios for life cycles of buildings and parts of buildings.

References

[CIA] Concrete Industry Alliance. 1998. Fact sheet. Crowthorne, Berkshire, GB: CIA. Available at www.bca.org.uk. Accessed 25 Mar 2003.

Miller A. 1999. Presentation at CIB/RILEM Congress. Gävle, S. a.miller@brighton.ac.uk.

7 Presentation and Communication of Results

The challenge of presenting and communicating life-cycle assessment (LCA) results depends on the goal and target group. The specificity in the building and construction (B/C) sector lies in the fact that building material and component combination (BMCC)–LCAs are used to calculate B/C–LCAs.

Besides the need for consistency of the LCA method applied, the presentation and communication of BMCC–LCA results should be clear for the user. It is critical that results be communicated in enough detail to make clear whether data for one product are truly comparable to data for another. Therefore, the presentation and communication should be standardised to a certain extent for specific goals and target groups.

The International Organization for Standardization (ISO) 14040 series contains requirements for the reporting of LCAs and the reviewing process. The requirements focus on the need to explain the LCA process (the way the research is carried out) and the transparency of the results. The way in which LCA results must be presented and communicated to target groups is not standardised (and cannot be in the general LCA framework because it is goal-specific). ISO 14040 does not provide the required detailed standards for presentation of BMCC–LCAs in B/C, and it can serve only as a basis for further standardisation.

Detailed LCA reports are suitable for LCA experts only, and often a further elaboration to obtain consistent data for inclusion in a BMCC–LCA database is required (such elaboration also is required to bring the report into conformity with ISO). Therefore, several attempts have been made to present BMCC–LCA results in the same way as other types of product information sheets, such as data sheets or labels. To support LCA practitioners and LCA data users in exchanging this type of information on a national and international level, there is a need for standardisation of these sheets or labels. ISO 14024/14025 (dealing with environmental labelling that may be based on LCA) form the basis for several labelling and declaration systems in European countries. Some of the systems are specific for B/C; others are not (although they do not exclude B/C).

ISO TC59 SC3 presents sustainable building standards and contains proposals for 5 standards:
1) Sustainable building terminology
2) Environmental declarations of building products

3) Sustainability indicators for the built environment and construction industry
4) Tools and methods for the assessment of the environmental sustainability of buildings
5) Tools and methods for the design of sustainable buildings.

It is not yet clear how far these standards will go toward fulfilling the required standardisation for clear data communication.

Easy-to-read and easy-to-use LCA-based product information is (at least for the B/C sector) still in its infancy. Developers and users of environmental product information systems in several European countries need to exchange information about existing systems. Therefore, the working group made an inventory of declarations and labelling systems used by the B/C sector (public systems only), shown in Table 7-1 (this does not deal with databases or specific LCA tools; these are listed in Appendix D). At the moment, we cannot provide more than an overview for information exchange purposes; it is too early to provide guidelines. Nevertheless, further standardisation on an international level is required for the near future. Communication about data consistency will be essential in any future discussions.

We recognise the following trends in product information systems:

- In countries where B/C-specific systems have been developed, such systems are preferred over general systems. Usually, guidance documents or manuals are available to describe building-specific issues concerning the LCA method to be applied. However, there are large differences in the levels of detail provided by such documents.
- In some countries, general systems are used by the B/C sector.
- Producers prefer Type III labels over Type I (Type I/III: see Definitions and Abbreviations, p 79). Only Type III labels are successful up to now.
- Not all systems include an external critical review.
- The data provided are either LCI or LCIA.

When we look further into the LCA methods and the formats of presentation, we conclude that information transfer between actors in the building chain, based on these product declarations, is very difficult. Information exchange between countries is even more problematic. Further discussions between actors (researchers, practitioners, producers, architects, commissioners, etc.) and countries will be important for the required further understanding, exchangeability, and (probably) standardisation of LCA information in B/C.

At the moment, declaration and labelling systems for BMCC-LCAs are very difficult or impossible to exchange between actors and countries. Further discussions, development, and harmonisation are essential for the successful use of LCA in the B/C sector.

Table 7-1 Declaration and labelling systems based on LCA used by the B/C sector in Europe

Country	Name of system	Type (ISO 14024/25)	Level in building chain	User and/owner of system	Goal and target group of data provided by label or declaration	LCA method	Critical review type	Remarks
Netherlands (Stichting NVTB Projecten 2000)	Environmental Relevant Product Information (MRPI).	Type III	Building materials, products, elements	User: producers (LCA carried out by an arbitrary LCA expert commissioned by producer). Owner: association of producers (NVTB).	Goal: data supply by BMCC LCAs. Target group: next actor in chain, either a producer or an architect.	Based on ISO and CML, extra rules specific for B/C on (described in manual).	External, by independent LCA expert, based on a standard review protocol provided by the MRPI system.	Presentation format (a kind of fact sheet) is standardised (described in manual).
UK (Howard et al. 1999)	BRE Environmental Profiles.	Type III	Building materials, products, elements	User: producers (LCA carried out by BRE). Owner: BRE.	Goal: Independent, level-playing-field information for producers to pass to clients and incorporate into various assessment tools.	Based on ISO and CML, extra rules specific for B/C on (described in report).	Certification required for external claims by companies. Generic information does not require certification	Presentation format is standardised.
Sweden[a]	Environmental product declaration (EPD).	Type III	Any product, including construction products.	User: consumers. Owner: Miljöstyrningsradet (Swedish EPA).	Product choice	ISO.	Public hearings to PSR product-specific requirements; EPDs reviewed by LCA experts.	Certified according to ISO standards on certification by a SWEDAC accredited body.
Denmark (Hansen et al. 2001)	EPD of building products.	Type III	Building products (materials and components delivered to building site).	Users: producers, consultants, and constructors. Owners: not yet decided.	Goal: provision of environmental data on building products, e.g., for incorporation in assessment tools.	Based on ISO and EDIP, use of aggregated effects, extra rules for B/C, e.g., for indoor climate.	External, by environmental certification bodies.	Presentation format is standardised. NB: the answers refer to preparatory decisions.
Norway	Environmental declaration of building materials.	Type II &III	Building materials, products, elements.	Users: producers, contractors, governmental organisations.	Target group: actors in the building industry being a producer, architect or contractor.	Based on ISO and CML (effect factors).	External, by the owner.	Presentation format (2 pages) is still in development.
Finland	RTS environmental declarations.	Type III	Building materials and products.	Users: producers. Owner: Building Information Foundation (RTS).	Designers.	Based on LCA (VTT method).	Basic system reviewed by a working group having representatives from production industry, designers, contractors, research institutes.	
Switzerland	SIA 493, Deklarationsrastervon Bauprodukten Schweizer Ingenieur- und Architektenverein, Zürich.	?	Building products.	Users: architects and planners. Owner: Producers, Swiss Federation of Architects and Engineers.	Goal: showing environmental preferences. Target group: architects and planners	No LCA method.	?	Presentation format is a SIA fact sheet
France	AFNOR XP P 01-010	Type III	Building products.	Users: industrial research centers, architects, project managers. Owner: AFNOR.	Goal: providing environmental information for product choice in construction projects.	Based on ISO 14040 and 14041 (part 1); partly based on ISO 14042 (part 2).	?	Presentation format is an environmental declaration data sheet (part 1), LCI based with defined categories.
Several European countries	Labelling systems for buildings and/or estates: UK: BREEAM Sweden: EcoEffect Norway: Ecoprofile Norway: ERCB (Environmental and Resource Commercial Buildings).	Types I & III	Building elements, buildings, estates.	BREEAM: BRE. EcoEffect: KTH. Ecoprofile: GRIP centre.	Target groups: building owners; sometimes designers. Goal: environmental classification of buildings.	BREEAM: BRE method. EcoEffect: EDIP. ERCB: LCA and GBC98.	Ecoprofile: certified people carry out the assessment.	Most often, standard presentations. Often includes indoor environment.

continued

Table 7-1 *continued*

Country	Name of system	Type (ISO 14024/25)	Level in building chain	User and/owner of system	Goal and target group of data provided by label or declaration	LCA method	Critical review type	Remarks
Several European countries and EU	Eco-labels, a.o.: EU: eco-label Dutch Milieukeur Nordic Swan German blue angel, etc	Type I	Any product.	User: producers. (LCA carried out by an arbitrary LCA expert commissioned by producer). Owner:EU	Goal: showing environmental preferences: Label is awarded to the best products within a product group. Target group: buyers of the product.	Specific requirements for label per product group are deducted by LCA; a producer must show by an LCA that the requirements are met.	External, by the owner.	System is hardly used by the B/C sector in NL and at EU level; the sector prefers Type III labeling.
Europe/CEPMC*	" CEPMC system" (no official system name; system is in testing stage).	Type III	Building materials and products.	User: producers. (LCA carried out by an arbitrary LCA expert commissioned by producer). Owner:association of producers (CEPMC).	Goal: data supply by BMCC-LCAs. Target group: next actor in chain, either a producer an or architect.	LCI data, based on ISO, no extra methodological rules.	None.	Presentation format (a kind of fact sheet) is standardised (described in a CEPMC guidance).

References

Hansen K, Fox M, Haugaard M, Krogh H, Skovsendem S. 2001. Environmental product declaration of building products (In Danish with English summary). Copenhagen, DK: Danish Environmental Protection Agency.

Howard N, Edwards S, Anderson J. 1999. BRE methodology for environmental profiles of construction materials, components and buildings. London, UK: BRE.

Stichting NVTB Projecten. 2000. Manual and background document for compiling environmentally relevant product information (MRPI), SNP-R98002 version 1.2, Driebergen, NL.

8 Relation of LCA in Building to Other Instruments

Besides life-cycle assessment (LCA), other analysis methods are related to environmental issues and chain analysis used in the building and construction (B/C) sector. In this chapter, 2 methods that are frequently used in the B/C sector will be discussed: life-cycle costing (LCC) and environmental impact assessment (EIA).

Life-Cycle Costing

Life-cycle assessment and life-cycle (or whole life) costing (LCC) share common features and aims, that is, they seek to assess impacts over the whole life of a building or structure and present the information in a manner that supports decision-making processes. The purpose of an LCC exercise is usually to aggregate total capital and operating costs of building systems and components over extended periods and then to present the figures as relative values that can easily be compared and assessed against alternatives. LCC does not explicitly deal with environmental impacts, although it can frequently be used to support environmentally sensitive construction solutions, especially where operating and/or maintenance costs are significant. A recent paper by the Building Research Establishment (BRE) details the many similarities and potential uses for combining LCC with LCA

(Edwards and Bartlett 2000). The key similarities are that both LCC and LCA use data on

- quantities of materials used,
- service life the materials could or will be used for,
- maintenance and operational impacts of using the products, and
- end-of-life proportions to recycling (and sale value) and disposal.

Such a combination has been realised in the German LEGOE (Hermann et al. 1998; Kohler and Hermann 2000; Kohler et al. 2001) and the Swiss OGIP (Kohler and Zimmerman 1996) LCA. (Information on these software tools can be found at www.legoe.de and www.ogip.ch, respectively.) It is particularly important to show the relations between design choices and resulting use costs (energy, maintenance, and operation cleaning). Cleaning costs are often higher than energy costs. There are still few inventories on cleaning products, while in many cases the massflow of cleaning products is much higher than the mass flow of the cleaned surface material.

The following are key differences between LCC and LCA:

- Conventional LCC methods do not consider the process of making a product; they are concerned with the market cost. LCA considers production.

- Life-cycle costs usually are discounted over time, whereas environmental impacts are not discounted.

The mechanics of LCC are well documented (Bull 1992; RICS 1992; Kirk and Dell'isola 1995), and the most typical approaches are those of net present value (NPV) and payback periods.

The NPV approach can be used to ascertain the relative financial merits of a number of design options over a specified period of time. The basic principle behind the method is that there is a time-value to money. Essentially, this reflects the fact that financial savings made in the distant future are of lesser value to the building owner than those made in the near future, considering the interest that could be accrued from investing the money up front and the effects of inflation over the study period. This process is known as 'discounting'. Selecting a discount rate is one of the most contested areas of LCC study from an environmental perspective because it also can be seen to be devaluing future resource consumption. Discounting distinguishes LCC from LCA, in which environmental impacts occurring in the future have exactly the same significance as those in the present. Discount rates in the order of 6% are generally considered acceptable in LCC. In order to perform this assessment, it is necessary to produce an inventory of yearly total costs, which include the capital, maintenance, operating, and energy costs of the systems.

In considering energy-saving measures, one of the simplest ways of considering the cost of additional capital investment is the payback period. The correct name for the technique is 'simple payback period', and it consists of defining the amount of time it will take to recover the initial investment in energy savings and dividing the initial installed cost by the annual energy cost savings. While simple payback is easy to compute,

its weakness is that it fails to factor in the time value of money, inflation, project lifetime, or operation and maintenance costs. To take these factors into account, a more detailed LCC analysis must be performed. Simple payback is useful for making ballpark estimates of how long it will take to recoup an initial investment.

Although these techniques are widely appreciated, the application of these techniques to real construction projects has achieved relatively little attention. Recent case studies of the use of LCA and LCC are provided in Edwards and Bartlett (2000). Data for capital costs of components and services are widely available, but obtaining data for operating costs can be a problem. Obtaining real data for building operative costs is a problem because they are widely considered commercially confidential.

Components and systems that reduce building environmental impacts are frequently associated with reduced life-cycle resource consumption and hence, usually, with reduced running costs. Some of these components or systems, however, may (but may not necessarily) incur additional capital costs. LCC should be used when a proposed option or solution is likely to affect both operating and capital costs.

When undertaking a life-cycle cost comparison, it is essential that like be compared with like; for example, if the costs of increasing wall insulation were considered, any additional structural components should be costed, as well as the possible reductions in the installation requirements. Just as an LCA would demonstrate an inventory of additional and avoided life-cycle environmental impacts, the life-cycle cost assessment should demonstrate an inventory of additional and avoided life-cycle costs. A life-cycle cost study is made more powerful if the proposed solutions are compared against a reference case, which refers to minimum building standards or typical practice.

An essential element of an LCC exercise is to consider the energy costs of all options, particularly where the proposed solution improves the energy efficiency of the building. Because energy prices can be expected to rise significantly during the life cycles of buildings, it is important that costs are assessed on the basis of national forecast statistics as well as current prices. Using a number of energy price scenarios will ascertain how sensitive the building system is to future price rises. An understanding of the sensitivity of the building to future price increases is potentially more important than trying to achieve an accurate forecast of energy prices.

Defining a period of study is a critical component of a combined LCC–LCA study, and will depend on a number of factors. The term 'life-cycle costing' actually is a bit of a misnomer because it is infrequent for the studied life to equate to the actual life. Assessing the operating life of a building is inherently difficult, and a building will usually reach the end of its prescribed definition of life before the end of its physical life. This disparity is known as 'obsolescence', and it occurs for a number of (mainly) functional and economic reasons but is also influenced by social and technical factors.

Even in the current economic climate, the economic performance of environmentally sound building systems is usually superior to that of more traditional systems over the life cycle of a building. Because we can expect the cost of consuming resources in the future to more accurately reflect the true environmental and social costs of their consumption, these arguments look set to become more convincing. Considering the scale of impact of the built environment on national and global scales, LCA and LCC should be seen as 2 common tools that will enable building designers and developers to recognise and demonstrate the benefits of environmentally sensitive construction.

Environmental Impact Assessment

For (large) construction works and major building sites, an EIA is often obligatory. EIAs usually focus on the environmental effects in the vicinity of the project. Even if a formal EIA procedure is not required, decision-making often requires information on local environmental burdens of a project, which can be seen as a kind of EIA.

Life-cycle assessment typically will be used in later phases than EIA: for more detailed choices on how to build, not on where to build. However, in some cases, the principles of sustainable building may extend to the earlier planning phases and an LCA becomes part of an EIA. Cases for this kind of concurring LCA–EIA are quite rare, but they do occur, for example, in the solar orientation of houses, the infrastructure in building plans, and the LCA of large infrastructural works. This chapter tries to deal with this overlap between EIA and LCA. A typical issue is to avoid double counting.

In such a case, local impacts can be assessed in greater detail than non-local impacts, for example, using specific distribution models for outputs and geographical data on sensitive areas. Sometimes it is easy to establish that local impacts are more (or less) important than non-local impacts of the same type (sensitive or nonsensitive areas). Some experience with such 'localised LCA' exists, but the members of the working group know no examples from the B/C sector.

Even if local impacts cannot be objectively discriminated from non-local impacts, it still might be wise to classify them separately because local impacts can be of special interest to decision-makers. Generally, local impacts will coincide with local political themes, which can be much more important for decision-makers than the abstract LCA results. To support the decision-making, one should provide all data that are relevant to the decision-maker and that clearly include specific, local impacts. Of course, if necessary, an objective LCA also should compare local impacts and non-local impacts, to show whether the non-local impacts are regarded as important by society as a whole.

This is especially relevant for a hindrance (land use, noise, smell). Usually, people involved in a building or construction project (e.g., spatial planners, architects, site managers) will already consider the hindrance at the site. Therefore, it usually is not common practice to include in a LCA the local hindrance at the site of the object under study. However, developments are going on to assess land use and the subsequent change of ecosystems in LCA, so this type of hindrance can be assessed in the same terms as other LCA issues.

A related topic is the development of regional characterisation models for LCA (especially work at the Technical University of Denmark; Potting 2000). Use of the model demands that for all major emissions in an LCA, the location is determined in more or less detail. This type of characterisation model implies that more data are needed on the locations where products are produced. Because many building materials are produced regionally, such data are readily available.

Mass flow accounting is another important method (Baccini and Brunner 1991; Behrensmeier and Bringezu 1995; Baccini and Bader 1996) that can be combined with LCA and EIA. The advantage is its link to macroeconomic data. This has been used in regional and urban analysis as well as for national building stocks. This leads to the following indicative table (Table 8-1).

Multi-Criteria Analysis

In construction, it is quite common to use integrated multiple criteria analysis for choices. The LCA results are then weighted against largely subjective other criteria with sometimes gross weighting factors (examples of other criteria: costs, safety, technical risks, aesthetics, hindrance [e.g., noise], local nature, other economical functions, flexibility, indoor climate, working conditions). Usually, only the costs are determined as precisely as the LCA. The LCA results must be weighted to a single score indicator to be usable in multi-criteria analyses.

Pressures on the time and the budget for an LCA can lead to oversimplifying LCA or to using full LCA but relying with too much confidence on existing databases and weighting factors. It will take time to develop working methods and models that are easy enough for designers to use but that do not oversimplify the decisions they have to make.

Table 8-1 Uses of LCA and EIA tools

Effects	On site (foreground process)	Not on site (background process)	Comments
Global effects only (global warming, ozone depletion, resource depletion)	LCA	LCA	
Regional effects (e.g., acidification, toxicity)	LCA or EIA, depending on specific study and theme.	LCA	Using regionalised characterisation models may improve description.
Local effects (e.g., noise, land use)	Do not use LCA; do use EIA.	Usually omitted in LCA, but should be included.	Onsite local effects tend to be highly weighted in local policy.

References

Baccini P, Bader H-P. 1996. Regionaler Stoffhaushalt. Berlin, DE: Spektrum Verlag.

Baccini P, Brunner PH. 1991. Metabolism of the anthroposphere. Berlin, DE: Springer-Verlag.

Behrensmeier R, Bringezu S. 1995. Zur Methodik der volkswirtschaftlichen Material-Intensitäts-Analyse: Ein quantitativer Vergleich des Umweltverbrauchs der bundesdeutschen Produktionssektoren. Wuppertal Papers Nr. 34.

Bull J, editor. 1992. Life cycle costing for construction. London, UK: Blackie Academic.

Edwards S, Bartlett E. 2000. BRE Digest 452: Whole life costing and life cycle assessment for sustainable building design. London, UK: CRC London. crc@construct.emap.co.uk

Hermann M, Kohler N, König H, Lützkendorf Th. 1998. CAAD system with integrated quantity surveying, energy calculation and LCA. Green Building Contest Conference (GBC '98); 1998 Oct; Vancouver, CAN.

Kirk SJ, Dell'isola A. 1995. Life cycle costing for design professionals. New York NY, USA: McGraw Hill.

Kohler N, Hermann M. 2000. Comprehensive and scalable method for LCA-cost and energy calculation. Maastricht. Conference Sustainable Building 2000; 2000 Oct 22–25; Maastricht, NL.

Kohler N, Hermann M, Lützkendorf Th, Schloesser D. June 2001. Integrated life cycle analysis. Submitted to Building Research and Information.

Kohler N, Zimmerman M. 1996. OGIP/DATO Optimierung der Gesamtanforderungen. Karlsruhe, DE: Universität Karlsruhe, Institut für Industrielle Bauproduktion (IFIB). www.ogip.ch. Accessed 26 Mar 2003.

Potting J. 2000. Spatial differentiation in life cycle impact assessment [PhD thesis]. Utrecht, DK: Utrecht Univ, Dept of Science, Technology & Society. ISBN 90-393-2326-7.

[R.I.C.S.] Royal Institute of Chartered Surveyors. 1992. Life cycle costing in buildings: Keynote lecture notes. Brighton, UK: RICS.

9 Conclusions and Outlook for the Future

Three main conclusions can be drawn from this report:

1) The performance requirements of building and construction (B/C) are the central subject of a B/C life-cycle assessment (LCA). Many identified issues and recommendations in this report are a consequence of this conclusion.

2) Building material and combination component (BMCC) LCAs (LCAs of materials, products, or components) is an essential part of B/C LCAs and standardisation of many LCA subjects is especially necessary on this level.

3) The success of LCA in B/C depends to a large extent on communication of BMCC data, of the performance concept and of usable results.

Based upon the conclusions, the workgroup identified the following to be important for the future:

- Further research for LCAs in B/C is required in the following areas:
  Harmonisation of BMCC–LCAs that will be used for B/C–LCAs. There should be a particular focus on harmonisation of system boundaries, modelling of maintenance, definition of end-of-life scenarios, and allocation.

- Related areas for research, either ongoing or needing further attention, that will impact upon the application of LCA to building materials include
  - the relation between durability and environmental effects (dynamic modelling),
  - the service life of BMCCs and service life planning of B/C and the development of consistent scenarios,
  - the clarification of the specificity and the interrelation between LCA and environmental impact assessment (EIA),
  - the ability of LCA to deal with indoor air quality (IAQ) and hazardous substances, and
  - the combination of LCA with life-cycle costing (LCC).

- Knowledge transfer and communication is required further to this state-of-the-art report, especially in the field of product declarations and labelling for BMCCs and B/C.

- Output results should be made accessible by
  - education,
  - environmental consciousness of all stakeholders in B/C, and
  - the presentation format of results.

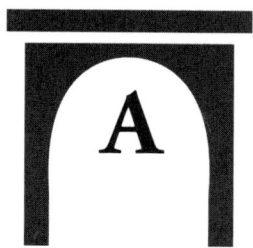

A SETAC Working Group on LCA in Building and Construction

Convenor/secretary
INTRON BV
Agnes Schuurmans
   Sittard, The Netherlands
   Also representative of the Dutch Association of LCA
   Consultants in the Building Industry

Chairwoman
Huntsman Polyurethanes
Shpresa Kotaji
   Everberg, Belgium

**Eidgenössische Technische Hochschule Zürich
(ETHZ)**
Annick Lalive d'Epinay, dipl.arch.ETH
   Group for Safety and Environmental Technology
   Zürich, Switzerland
   Since April 2000: Basler & Hofmann
   Ingenieure und Planer AG
   Zürich, Switzerland

Ecole Polytechnique Fédérale de Lausanne (EPFL)
Jean-Bernard Gay
   Laboratoire d'Energie Soliare et de Physique du
   Bâtiment
   Lausanne, Switzerland
   Corresponding member

Centre for Sustainable Construction
Building Research Establishment (BRE)
Suzy Edwards
   Garston
   Watford, Herts, United Kingdom

Atlantic Consulting
Eric Johnson
   Belgravia Workshops
   London, United Kingdom
   Corresponding member

Rijkswaterstaat (RWS)
Joris Broers
   Delft, The Netherlands

Institut für INDUSTRIELLE OKOLOGIE
Dr. Francois Schneider
   Tor zum Landhaus
   St. Pölten, Austria

Universität Stuttgart/IKP
Johannes Kreibig
   Abt. Ganzheitliche Bilanzierung
   Stuttgart, Germany

D.S.N.
Elisabeth Payeux
   Paris, France
   Corresponding member

Royal Institute of Technology, (KTH)
Dep. of Building Sciences
Div. of Building Materials
Mathias Borg, M.Sc.
   Stockholm, Sweden

**Belgian Building Research Institute
(CSTC-BBRI-WTCB)**
Jan Desmyter
(replaced by Johan van Dessel in 1st meeting)
   Brussels, Belgium

Life-Cycle Assessment in Building and Construction. Shpresa Kotaji et al., editors.
©2003 Society of Environmental Toxicology and Chemistry (SETAC). ISBN 1-880661-59-7

Centre d'Energetique, Ecole des Mines de Paris
Mr Bruno Peuportier
Paris, France
Corresponding member

Laborataire Central des Ponts E. Chaussèes (LCPC)
SEG.DNNCER
Gaétana Quaranta
Nantes, France
Corresponding member

DTI (Danish Technological Institute)
M. Haugaard
Taastrup, Denmark

Together with:

SBI
Klaus Hansen/ Hanne Krogh
Horsholm, Denmark

IPU/LCC-DTU
Mrs ir J. Potting
Lyngby, Denmark
Corresponding member

SBI
Joakim Widman
Stockholm, Sweden

IVAM Environmental Research
Jaap Kortman
Amsterdam, The Netherlands

Randa Group
Environmental Consultancy
Silvia Ayuso/Marta Vallès
Barcelona, Spain

TNO B/CResearch
Mrs. Adrie de Groot-van Dam
Delft, The Netherlands

Norwegian Building Institute
Building Technology Department
Ph.D Trine Dyrstad Pettersen
Oslo, Norway

fstfold Research Foundation (STO)
C.H. Borchsenius
Fredrikstad, Norway
Corresponding member

Industry Canada
Peter Mikelsons, M.B.A., P.Eng
Ottawa, Ontario, Canada
Corresponding member

VTT Building technology
mrs. Dr.Tarja Häkkinen
VTT, Finland

Umweltbundesambt Berlin
Dr. Hans-Hermann Eggers
Berlin, Germany

Chalmers University of Technology
Thomas Bjorklund
Technical Environmental Planning
Goteborg, Sweden
Corresponding member

Chalmers Industriteknik
Tomas Ekvall
Goteborg, Sweden
Corresponding member

Centre Scientifique et Technique du Batiment (CSTB)
Jean-François Le Teno
Grenoble. France

Institute for Building Materials (IBWK)
Christine Haag
Zürich, Switzerland
Corresponding member

British Cement Association (BCA)
Leslie J. Parrott, BSc (Eng), PhD
Crowthorne, Berks, United Kingdom

Universitat Polytecnica de Catalunya
Dr. Ignasi CASANOVA
Barcelona, Spain
Corresponding member

University of Salford
Mr Bryn Golton
Department of Environmental Resources
Salford, United Kingdom

KTH – Byggd Miljö
Dr Mauritz Glaumann
Gävle, Sweden
Corresponding member

Mrs Ginette Dupuy Gouin
Montreal, Quebec, Canada
Corresponding member

Athena Sustainable Materials Institute
Mr Jamie K Meil
Ottawa, Ontario, Canada
Corresponding member

NIST, Building and Fire Research Laboratory
Barbara Lippiatt
Office of Applied Economics
Gaithersburg, Maryland, USA

University of Brighton
Dr Andrew Miller
Brighton, United Kingdom

Kungl Tekniska Högskolan (KTH)
Professor Kai Ödeen
Division of Building Materials
Stockholm, Sweden
Corresponding member

Wayne B Trusty & Associates Ltd
Mr Wayne B Trusty
Merrickville, Ontario, Canada

John Emery Geotechnical Engineering Ltd
Dr J Emery
Merrickville, Ontario, Canada
Corresponding member

Institute of Industrial Science, University of Tokyo
Toshiharu Ikaga, Associate Professor
Department of Architecture and Civil Engineering
Tokyo, Japan
Corresponding member

Technische Universitaet Berlin
Dipl.-Ing. Andreas Ciroth
Institut fuer Technischen Umweltschutz
Abfallvermeidung und Sekundaerrohstoffwirtschaft
Berlin, Germany

EMPA Duebendorf
Frank Werner/115
Duebendorf, Germany

Ecobilan
Pascale Jean
Nanterre, France
Corrresponding member

The Steel Construction Institute
Tony Birtles
Ascot, Berkshire, United Kingdom

Carillilon Building
Andrew Horseley
Wolverhampton, West Midlands, United Kingdom

Royal Institute of Technology, (KTH)
Jacob Paulson
Department of Building Sciences
Division of Building Materials
Stockholm, Sweden

Ingenieurbüro Trinius
Wolfram Trinius
Hamburg, Germany

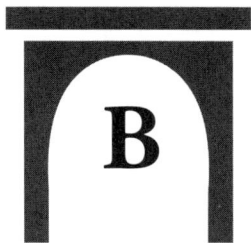

B Sensitivity of Transport

The study of the environmental impacts of building materials throughout their life cycle should include consideration of transportation within and between each stage of the life cycle.

The energy consumption for transportation will vary greatly for different materials and even for similar materials using different sources of raw materials or different networks for distribution. Published figures for transportation are inconsistent, probably because they include different things. However, it generally is not possible to determine what has been included in any given figures for the implications of transportation.

Transport energy (published data)

Canada	1.18 MJ/tonne/km
Denmark	1.44 MJ/tonne/km
Sweden (Long distance)	1.0 MJ/tonne/km
Sweden (Short distance)	2.7 MJ/tonne/km
UK	4.5 MJ/tonne/km
U.S.	2.13 MJ/tonne/km

It is evident that fuel consumption is dependent upon mode of transport, efficiency of vehicles, efficiency of loading, and efficiency of distribution network. The significance of transportation within the impacts of given materials will vary depending upon the amount of processing required by those materials.

In order to improve our evaluation of the environmental impacts of transportation and to ensure that research efforts are directed towards the evaluation of the most significant transportation impacts, we must undertake a sensitivity analysis of the factors that affect transportation impacts.

The proposal, therefore, is to develop a checklist of the factors affecting the environmental impacts of transportation of building materials and to undertake a sensitivity analysis of the impacts of these factors.

Some factors affecting transportation fuel consumption to take into account will include
- type of vehicle,
- level of maintenance,
- traffic,
- route restrictions,
- return journey,
- urban or rural journey,
- distance, and
- loading.

It is relevant to notice that life-cycle transportation energy is affected by 3 main elements: the energy consumed by the transportation, the energy linked to the manufacture and maintenance of the vehicle, and the energy linked to the construction and maintenance of the roads (or rails). The contribution to the total transportation life-cycle energy profile accounts for 61%, 32%, and 7% respectively.

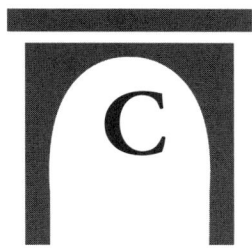

C Representativity for Local Materials

Inventory: Local Materials and Local Process Data

For low-grade heavyweight building materials, markets are regional because prices prohibit long-distance transport. Even in small countries like the Netherlands and Denmark, materials such as filling sand and road base materials have regional markets. For higher-grade materials, longer transport distances are common. In larger countries like the UK or Canada, markets for products such as concrete aggregates and concrete products are local too. Asphalt and concrete mortar are transported as a hot or reactive mixtures and cannot be transported over large distances.

It is generally acknowledged that one should not use world averages for regionally produced products. The general principle should be that one should use the average of the environmental profile on the market from which a certain product will be bought. For example, because there is a European market for electricity, it seems to be appropriate to use the European electricity mix. For many (raw) materials and basic products, markets are global, so we need global averages that include, for example, Asian data.

For low-grade heavyweight building materials, markets are regional, so regional data should be used. However, these data usually are not available.

Currently, it is quite common to use national averages, even in cases where such data are known to be inaccurate. A more exact interim procedure is sometimes used: A regional mix of production technologies is used (e.g., sea dredging, pit dredging, quarry), but the life-cycle inventory (LCI) data for these production technologies are national or European, assuming that these data do not vary too much amongst producers.

Inventory: Transport

For materials that involve little processing other than dredging or digging and transport—for example, gravel, sand, rock, and clay—transport is the dominating term in the impact assessment (except for the 'land use' impact category). For wood, transport is significant if the embodied (solar) energy is not included. But also for higher-grade materials like (asphalt) concrete, transport may make a significant contribution to energy use and land use, which can be quite important aspects in a life-cycle assessment (LCA) of building materials. For products like plastic and steel, which have a much higher embodied energy, the production will usually dominate the transport energy.

It appears that generally default transport distances are used in building LCAs. However, very few LCAs deal with materials like sand, and in these studies, the transport usually is taken into account explicitly. Because of their high contribution to the environmental profile, especially when dealing with low-grade materials, we suggest that such standard distances be replaced by specific estimated distances for the project under study as soon as possible.

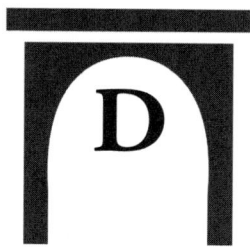

D Survey of LCA Literature for Building and Construction

Data and information related to life-cycle assessment (LCA) in building and construction (B/C) is available worldwide. Here we provide you with references to

- Books and reports,
- Product life-cycle studies,
- Websites, and
- LCA softwares.

Books and Reports

Alef K, Fiedler H, Hauthal W. 1997. Eco-Informa 97: Information and communication in environmental and health issues. 12, Eco-Informa Press, Ispra (I), 6.10.–9.10.

Anink D, en Mak J. 1993. Milieueffecten van kozijnreparatie en kozijnvervanging, Eindrapport. (Window frames en repair). Gouda: Woonenergie.

Baccini P. 1998. Welche Ressourcen stecken in den Bauwerken unserer Siedlungen? Eine Einführung zur Wahl der Begriffe und der Untersuchungsmethoden. In: Lichtensteiger Th (Hrsg): Ressourcen im Bau. ETH Zürich, EAWAG. Hochschulverlag AG an der ETH Zürich, vdf. Zürich. 3-7281-2638-1.

Baccini P, Brunner P. 1991. Metabolism of the anthroposphere. Berlin: Springer Verlag.

Baccini P, Bader HP. 1996. Regionaler Stoffhaushalt – Erfassung, Bewertung und Steuerung. Heidelberg, Berlin, Oxford: Spektrum Akademischer Verlag. 3-86025-235-6.

Baré F, Bosshard C, Contich P, Jegher G, Karlaganis G, Leumann P, Loup PA, et al. 1995. BUWAL, Empfehlungen und Grundlagen für Malerarbeiten.Bern: Bundesamt für Umwelt, Wald und Landschaft. Mitteilungen zum Gewässerschutz NR. 16.

Becalli G, Cellura M, Becalli M. 1996. Life Cycle Assessment of building materials - a comparison by electre methodology. 7th International Conference on the Durability of Building Materials and Components. Stockholm, Sweden: Royal Institute of Technology.

Binz A, Frischknecht R, Gilgen D, Gugerli H, Lehmann G. 2000. Ökologische Sanierung von Bürobauten. Beitrag im Rahmen des IEA CBS Annex 31, zu beziehen bei: ZEN, EMPA Dübendorf. Zürich.

Blume J, Simm A. 1994. Sichtanalyse des im Landkreis Böblingen deponierten Gewerbe- und Baustellenabfalls. Abfallwirtschaftsjournal, EF-Verlag für Energie- und Umwelttechnik GmbH, 6:89–98.

Bolderman R, Bijen JM, Schuurmans A, et al. 1996. De milieubelasting van buismaterialen. Addendum: LCA van buizen van nodulair gietijzer in vergelijking met gewapend beton.

(sewer pipes, steel en reinforcec concrete). Woerden: VPB rioleringstechniek.

Boonstra Ch, Kohler N, Pagani R, Peuportier B. 1997. REGENER final reports, C.E.C. DG XII contract n° RENA CT94-0033, Ecole des Mines de Paris.

Braunschweig A. 1991. Ökologischer Vergleich der Transporte vn Kies und Aushub von und nach Hüntwangen mit Bahn oder LkW. St. Gallen: Bericht AGÖK.

Bretz R, Fankhauser P. 1996. Screening LCA for Large Numbers of Products: Estimation Tools to Fill Data Gaps. Grenzach: Ciba-Geigy Limited.

Burkhardt P. 1999. E-Top Rating. In: A. Lalive d'Epinay und D. Quack (Hrsg.) Ökologische Bewertung von Gebäuden zwischen Forschung und Praxis. Begleitende Unterlagen zum 11. Diskussionsforum Ökobilanzen vom 1 November 1999. Zürich: ETH Zürich, Gruppe für Sicherheit und Umweltschutz, Laboratorium für Technische Chemie.

Christophersen E, Dinesen J, Nielsen P.1993. Life Cycle based Building Design. Proceedings 3rd Symposium Building Physics in the Nordic Countries. Copenhagen.

Coutalides R. 1993. Ökoschirmbild Stahlrain. Interner Bericht Metron Raumplanung AG, Brugg.

Deliege EJM, Nijdam DS en Veen, van W. 1997. LCA AVI-vliegas. (MSW fly ash), Deventer: Tauw Milieu B.V.Projectnummer 3601064.

Dinesen J, Krogh H, Traberg-Borup S. 1997. Life-Cycle-Based Building Design. Denmark: SBI-Report 279.

Dinesen J, Traberg-Borup S. 1994. An Energy Life Cycle Assessment Model for Building Design. Danish Building Research Institute, Energy and Indoor Climate Division. Paper No. 13.

DHV AIB B.V. 1998. LCA Dakproducten RBB, Teewenpan en KDN ; eindrapport ; definitief (roofing tiles), Amersfoort. PMC-7904.

Eaton K, Amato A. 1996. Assessing the environmental impact of structural flexibility. In: IABSE, 15th Congress. Vol I. Kopenhagen: International Association for Bridge and Structural Engineering. p 38-394.

Erlandsson M, Mingarini K, Nilvér K, Sundberg K, Ödéen K.1994. Life Cycle Assessment of Building Components. Department of Building Materials, The Royal Institute of Technology. Swedish Waste Research Council. Annual Report.

European Commission. 1997. Directorate General XII for Science, Research and Development, Programme APAS; European methodology for the evaluation of environmental impact of buildings: Introduction to Life Cycle Analysis of Buildings, Final Report Part 1, European Commission.

European Commission. 1997. Directorate General XII for Science, Research and Development, Programme APAS, European methodology for the evaluation of environmental impact of buildings: Applications of the Life Cycle Analysis to Buildings, Final Report Part 2, European Commission.

European Commission. 1997.Directorate General XII for Science, Research and Development, Programme APAS, European methodology for the evaluation of environmental impact of buildings: The Integration of Environmental Assessment in the Building Design Process, Final Report Part 3, European Commission.

Eyrerer P. 1996. Ganzheitliche Bilanzierung – Werkzeug zum Planen und Wirtschaften in Kreisläufen. Heidelberg: Springer Verlag.

Eyrerer P, Reinhardt HW. 1999. Ökologische Bilanzierung von Baustoffen und Gebäuden – Wege zu einer ganzheitlichen Betrachtung. Zürich: Birkhäuser Verlag.

Fawer M. 1997. Life Cycle Inventories for the Production of Sodium Silicates, Dubendorf: EMPA, Bericht nr. 241.

Fluitman A. en Lange VPA de. 1996. Vergelijking van de milieu-effecten van drie betonnen verdiepingsvloeren. (concrete floors), Amsterdam CREM. CREM rapport nr 95.107.

Fossdal S. 1995. Energy Consumption and Environmental Impact of Buildings in Norway: Life Cycle Assessment, International Energy Agency. Energy Conservation News, Issue 22.

Fossdal S, Edvardsen KI. 1995. Energy consumption and environmental impact of buildings. In: Buildings Research and Information Vol. 23(4):221-226, E.&F.N. Spon.

Fraanje P, et al. 1992. Milieubelasting van twee aanbruggen: Een pilot study. (part of bridges) Amsterdam: Interfacultaire Vakgroep Milieukunde Universiteit van Amsterdam. IVAM Onderzoeksreeks ; 57. ISBN 9072011201.

Friedli R. 1998. OGIP Ein Bau-Planer-Werkzeug zur Beurteilung der Ressourcen Kosten/Energie/Umwelt. in: M. Zimmermann & H. Bertschinger (Hrsg.): 10. Schweizerisches Statusseminar. EMPA-KWH, Bundesamt für Energiewirtschaft, ETH Zürich.

Frühwald A, Scharai-Rad J, Wegener G, Zimmer B. 1997. Informationsdienst Holz – Ökobilanzen Forst Fakten lesen, verstehen und Handeln. München: Deutsche Gesellschaft für Holzforschung.

Frühwald A, Scharai-Rad J, Wegener G, Zimmer B. 1997. Informationsdienst Holz – Erstellung von Ökobilanzen für die Forst – und Holzwirtschaft. München: Deutsche Gesellschaft für Holzforschung.

Gay F-B, Homem de Freitas J, Ospelt Ch, Rittmeyer P, Sindayigaya O. 1996. Toward a Sustainability Indicator for Buildings (Workshop on Future of the Cities September 16-17, 1996 MIT, Cambridge), LESO-EPFL Lausanne, Lausanne.

Gerber D, Haas A. 2000. Vergleichende Ökobilanz von Niedrigenergiehäusern. Bern: Gefördert im Rahmen des BfE-Programmes Solararchitektur und Tageslichtnutzung, Projektnummer 21 082, ENET.

Gorgolewski M, Eaton K. 1995. Environmental Impacts of Steel Piling Steel Construction Institute (UK). Presented at Sustainable Steel Conference; Orlando.

Gorree M, en Kleijn R. 1996. Screening-LCA voor de verwijdering van baggerspecie. (sludge) Leiden: Centrum voor Milieukunde, CML, Rijksuniversiteit Leiden. Rapport 129.

Gugerli H. 1999. Ökobilanz des Bürogebäudes der UBS in Suglio, Tessin. Zürich: nterner Bericht, INTEP, Amstein und Walthert Ingenieure und Planer.

Häkkinen T, Ahola P, Vanhatalo L, Merra A. 1999. Environmental impact of coated exterior wooden cladding. Finnish version published in VTT Research Notes, 834). Espoo. English version in www.vtt.fi/rte/projects/environ/enviro_prj_paints.html.

HäkkinenT, Kronlöf A. 1994. Environmental assessment of building materials. VTT Research Notes 1591. Espoo. 61 p. + app. 25 p. (In Finnish). English shortened version in VTT Research Notes 1590. Espoo 38 p.

Häkkinen T, Mäkelä K. 1996. Environmental adaptation of concrete. Environmental impact of concrete and asphalt pavements. Espoo: VTT Research Notes 1752. 61 p. + app. 32 p.

Häkkinen T, et al. 1997. Environmental declarations of building products. Espoo: VTT Research Notes 1836. 138 p. + app. 10 p. (In Finnish).

Häkkinen T, Saari M, Vares S, Vesikari E. 1999. Design of eco-efficient building. 91 p. Rakennustieto. Helsinki. (In Finnish).

Hendriks NA, Meijden CW, van der en Oomen W. 1996. Het leven van alle daken (HLD): Eindrapport. (roofs), Gorinchem: BDA Dakadvies. Opdrachtnummer: 94-B-0059.

Hendriksen LJAM, en Eggels PG. 1994. Milieubelasting door comfortinstallaties in woningen. Pilot studie bruikbaarheid LCA methode. (installations), Rotterda: Instituut voor studie en stimulering van onderzoek op het gebied van gebouwinstallaties, ISSO. rapport 32.01.

Hoefnagels F, en Lange V de. 1992. De milieubelasting van vier verfsystemen : Standaard Oplosmiddelhoudende Alkydverf, Watergedragen Acrylaatdispersieverf, High Solid verf, Solvent Free verf. (paints). Amsterdam: Consultancy and Research for Environmental Management, CREM.

Hoefnagels F, en Lange V de. 1993. De milieubelasting van houten en betonnen dwarsliggers (wooden en concrete layers for railways). In: opdracht van Bureau Milieu, Infrabeheer, NV Nederlandse Spoorwegen, Amsterdam: Consultancy and Research for Environmental Management, CREM. 92.019a.

Hoefnagels FET, en Ratering MT. 1994. Milieugerichte levenscyclusanalyse (LCA) van EPS en steenwol toegepast als isolatie van begane grondvl9oeren. (EPS insulatin in floors), Amsterdam: Consultancy and Research for Environmental Management, CREM. 94068.

Hoefnagels F, Kortman J, en Lindeijer E. 1992. Minimalisering van de milieubelasting van buitenkozijnen in de woningbouw. (window frames), Amsterdam: Interfacultaire Vakgroep Milieukunde (UvA) IVAM onderzoeksreeks. 54. ISBN 9072011171.

Hofmann M, Meier H-P, von Willisen FK. 1995. Werkstoffinnovationen im schweizerischen Umfeld. Bern : Schweizerischer Wissenschaftsrat.

Hoog J op 't. 1993. Levenscyclusanalyse van ftovoltaische systemen: Milieu-effecten van wieg tot graf. Deel1: De Methodologie. Deel 2: Een voorbeeldstudie; fotovoltaische systemen. Afstudeerrapport. Eindhoven: Technische Universiteiten te Eindhoven, TUE.

Howards N, Edwards S, Anderson J. 1999. BRE Methodology for Environmental Profiles of Construction Materials, Components and Buildings. Watford.

Infras. 1996. Nachhaltigkeit des Bauens in der Schweiz. Zürich: Koordinationsgruppe des Bundes für Energie- und Ökobilanzen. ENET Nr. 30513.

INTRON. 1994. Milieuprofiel van hpl compact plaatmateriaal (HPL cladding panels). Trespa Hoechst Holland N.V, 94071 G38120.

INTRON. 1995. Indicatieve levenscyclusanalyse van PVC en gres rioolbuizen in vergelijking met beton. (sewer pipes PVC and clay compared to concrete). INTRON. VPB 95195 M701150,

INTRON. 1996. Globale levenscyclusanalyse verbijzondering knooppunt Ridderkerk (light line). Rijkswaterstaat Directie Zuid-Holland, 96186 M708010.

INTRON. 1996. Levenscyclusanalyse woningscheidende vloerconstructies met lewis zwaluwstaartplaten. (steel floor plates). Reppel, 96371 M706810.

INTRON. 1996. Milieufacetten van beton. Beschrijving database inputgegevens voor LCA's van betonnen producten in de woning en utiliteitsbouw. (LCA concrete database), Betonplatform. 95406 G700260.

INTRON. 1997. LCA voor prodakproducten (sandwichpanels and roof panels). prodak, 97109 M711330.

INTRON. 1997. Milieugerichte levenscyclusanalyse van woningscheidende vloerconstructies met lewis zwaluwstaartplaten: Samenvatting. (steel floor plates), Reppel B.V, 97235 M712360.

INTRON. 1997. Milieugerichte levenscyclusanalyse van betonnen wanden met menggranulaat. Een vergelijking vanuit meerdere invalshoeken. VNC en Heembeton B.V, 97262 M712380.

INTRON. 1997. Vergelijking van milieueffecten van de winning van ophoogzand uit zee met de winning uit binnenlandse voorraden, NVTB. 96349 M708610, M711930.

INTRON. 1998. Eco-profiles of high pressure decorative laminate (HPL) according to en 438.1 and its elements. European data of the ICDLI. Summary Report, ICDLI, 97318 M707900, april.

INTRON. 1998. Een energieanalyse van varianten voor aanpassing van de RW16/13. Duurzaam bouwen Rijksweg 16/13, grondstoffen en energie (raw materials and energy analysis for road RW16/13). Bouwdienst Rijkswaterstaat, 980548, M715240.

INTRON. 1998. LCA van betonnen tegels (concrete pavement tiles). Brancheverenigingsgegevens van Teban. BFBN, 980152 M712270.

INTRON. 1998. LCA van kalksteenbeton. De milieueffecten van de recycling van kalksteenbeton in klinkerbereiding. (concrete with limestone). ENCI N.V, 980388 M715210.

INTRON. 1998. LCA voor kalkzandsteen, een milieuanalyse van de productie, transport carbonatatie en recycling van stenen en blokken, elementen en gevelstenen (sandlime brick). RCK, 980379a M711520.

INTRON. 1998. Levenscyclusanalyse van kunststof gevelelementen. Een milieuanalyse voor Stichting Recycling VKG in het kader van de haalbaarheidsstudie, productgerichte milieuzorg (PVC window frames). Witteveen en Bos, raadgevende ingenieurs B.V, 980382 M711600.

INTRON. 1998. Milieu-en energieaspecten Betonkringloop gesloten (closed concrete chain). VNC, 980288 M713870.

INTRON. 1998. Milieugerichte Levenscyclusanalyse van Rockwool pijpschalen (rockwool pipe shales). Rockwool Lapinus b.v, 97179 M711760.

INTRON. 1999. LCA van afdichtingsmaterialen. een milieuanalyse van Cocoband 15/3 en 40/20 (sealing tape). Cocon Arkel. 990018a M716360.

INTRON. 1999. LCA van betonnen vloeren van Bevlon : Gegevens voor toetsing MRPI (concrete floors). BEVLON, 990325 A802520.

INTRON. 1999. LCA van Casco's in beton Een vergelijking tussen een Heembeton 949 casco en tunnelgietbouw, voor intern gebruik (concrete walls). Heembeton B.V, 990068 M716430.

INTRON. 1999. LCA van compressieband voor coegafdichting. een milieuanalyse van MAVOTEX 300 compressieband 15/3 (compression tape). Mavotrans, 990066a M716360.

INTRON. 1999. LCA van compressieband voor voegafdichting. Een milieuanalyse van ILLMOD 600compressieband 15/3en 40/17-32 (compresssion tape). Mavotrans 990017a M716360.

INTRON. 1999. LCA van Rockwool spouwplaat 433 en 433 duo - gegevens voor toetsing MRPI (rockwool insulation material). Rockwool Lapinus, 9900071 M715940.

INTRON. 1999. Structural Sealant Glazing SSG beoordeeld op milieuaspecten met een indicatieve levenscyclusanalyse (LCA). Keers Konstructiewerken, 9900018 M716030.

INTRON. 2000. De energieanalyse van inrichtingsvarianten voor aanpassing van de RW12. Duurzaam bouwen Rijksweg 12, grondstoffen en energie (raw materials and energy analysis for road RW12). Bouwdienst Rijkswaterstaat, 9900086, A902080.

INTRON. 2000. LCA van Casco's in beton (concrete casco's). Heembeton bv, 990068 716430.

INTRON, LCA-studies for Environmental Relevant Product Information (MRPI) for:
· YTONG, cellular concrete
· Hebel, cellular concrete
· ENCI, blast furnace slag cement
· VOBN, ready-mixed concrete
· KNB, brick
· RCK, sandlime brick
· VKG, PVC window frames
· BFLL, plastic roofing windows
· Vekudak, plastic roofing material
· Ubbink, plastic gutters
· NVPU, PUR insulation material
· Rockwool, rockwool insulation material
· Isover, glass wool insulation material
· Exiba, XPD insulation material
· Reppel, steel floor plates
· TRESPA, cladding panels
· Stidawa, plastic building foil
· Lafarge, roofing tiles

IP Bau, Bauabfälle, Teil des Stoffkreislaufs. Bundesamt für Konjunkturfragen, Bern, 1993.

Jacobs F, von Arx U, Spanka G, Mäder U, Chudacoff M, Bauchemie: Umwelt kontra Bautechnik, in Wildegg, S, Materialtech-nologische Aus- und Weiterbildung. Vol. 5141, Seiten: 107, TFB, S. W, Technische Forschungs- und Beratungsstelle der Schweizerischen Zementindustrie TFB, Wildegg, 1995.

Klöpffer W. Prinzipielle Möglichkeiten der Ökobilanzierung von Baustoffen, Persönliche Mitteilung, 1997.

Koch P, Seiler B, Ott W. 1998. Funktionale Einheit und Systemgrenzen bei Ökobilanzen im Bauwesen. Econcept, Gruppe für Sicherheit und Umweltschutz, Laboratorium für Technische Chemie, Universitätsstrasse 31, 8092 Zürich, February, 1998.

Kohler N. 1994. Energie- und Stoffflussbilanzen von Gebäuden während ihrer Lebensdauer. Bericht Bundesamt für Energiewirtschaft, Bundesamt für Umwelt, Wald und Landschaft, Amt für Bundesbauten, Bundesamt für Konjunkturfragen, Karlsruhe, 1994.

Kohler N. 1996. Energie- und Stoffflussbilanzen von Gebäuden: Stand der Forschung und Perspektiven in: ifib-Veröffentlichungen°Nr. 9, 1996.

Kohler N. 1999. Lebenszyklusbezogene Bewertung von Gebäuden, Methoden und Vergleiche. In: Deutsches Architekturmuseum und M. Volz (Hrsg.): Die ökologische Herausforderung. Ernst Wasmuth Verlag Thübingen, Berlin, 1999. 3-8030-0193-5.

Kohler N, Barth B, Bieber H, Eiermann O, Haida A, Heitz S, Hermann M, Klingele M, Koch M, Kukul E, Holliger M, Frischknecht R, Stritz A, Weibel Th, Goeggel HP, Sachs S. 1996b. Schlussbericht BEW Forschungsprojekt OGIP/DATO – Optimierung von Gesamtenergieverbrauch, Umweltbelastung und Baukosten. ifib Universität Karlsruhe, Englerstrasse 7, 76128 Karlsruhe, 1996.

Kohler N, Holliger M, Lützkendorf Th. 1992. Regeln zur Datenerfassung für Energie- und Stoffflussanalysen, Handbuch, Lausanne 1992.

Kohler N, Lützkendorf Th. 1990. Energie- und Schadstoffbilanzen bei Niedrigenergiehäusern, Teil 1, Grundlagen, Zwischenbericht, EPFL Lausanne, 1990.

Kohler N, Lützkendorf Th, Holliger M. 1991. Energie- und Schadstoffbilanzen im Bauwesen, EPFL Lausanne, Hochschule für Architektur und Bauwesen Weimar, Lausanne 1991.

Kohler N, Lützkendorf Th, Holliger M. 1992. Methodische Grundlagen für Energie- und Stoffflussanalysen - Handbuch, Diskussionsbeitrag zum BEW-Projekt Energie- und Stoffbilanzen von Bauteilen und Gebäuden, Koordinationsgruppe des Bundes für Energie- und Oekobilanzen, 1992.

Kohler N et al. 1995. Baustoffdaten - Oekoinventare, TH Karlsruhe, ETH Zürich, Karlsruhe/Weimar/Zürich 1995.

Kohler N et al. 1996. KOBEK - Methode zur kombinierten Berechnung von Energiebedarf, Umweltbelastung und Baukosten in frühen Planungsstadien, Schlussbericht Deutsche Bundesstiftung Umwelt, Karlsruhe 1996.

Kollbrunner R, Müller W. Proposition and ecological Assessment of alternative air conditioning systems. Diplomarbeit, Institut für Energietechnik, ETH, 111 + Annex Seiten, Zürich, 1994.

Köllner Th. 1999. Life Cycle Impact characterisation for land use: assessing the impacts on the regional plant species pool diversity. In: SETAC (Hrsg.): Quality of Life and Environment in Cultured Landscapes, 9th Annual Meeting of SETAC Europe, Leipzig, Deutschland, 25–29. Mai, 1999.

Kortman JGM. en Lim RG. Minimalisering van de milieubelasting van niet-dragende binnenwanden in de woningbouw. (interior non load bearing walls), Amsterdam : Interfacultaire Vakgroep Milieukunde (UvA), 1993, IVAM onderzoeksreeks ; 65, ISBN: 9072011279.

Kreissig J, Baitz M (IKP, University of Stuttgart), Kümmel J (IWB, University of Stuttgart). Life Cycle Engineering for the Architect – Environmental rating of structural components with reference to the entire system.

Kreissig J, Baitz M, Betz M, Eyerer P, Kümmel J, Reinhardt H: Leitfaden zur Erstellung von Sachbilanzen in Betrieben der Steine-Erden_Industrie, Bundesverband Steine + Erden, Frankfurt, 1997.

Kreissig J, Baitz M, Betz M, Eyerer P, Straub W. Ganzheitliche Bilanzierung von Fenstern uns Fassaden, IKP, Uni Stuttgart, 1998.

Krusche P, Krusche M, Althaus D, Gabriel I. Ökologisches Bauen. Umweltbundesamt, Wiesbaden, Berlin, 1982.

Künniger T, Richter K. 1997. Ökologischer Vergleich von Freileitungsmasten aus imprägniertem Holz, armiertem Beton und korrosionsgeschütztem Stahl. Eidg.

Materialprüfungs-und Forschungsanstalt (EMPA), Dübendorf.

Künniger T, Richter K. 2000. Ökobilanzen von Konstruktionen im Garten- und Landschaftsbau. Eidg. Materialprüfungs- und Forschungsanstalt (EMPA), Dübendorf.

Lalive d'Epinay A. 2000. Die Umweltverträglichkeit als eine Determinante des architektonischen Entwurfs. Dissertation ETH Nr. 13610, Abteilung für Umweltnaturwissenschaften, ETH Zürich, März, 2000.

Lalive d'Epinay A, Müller A, Pulli R. 1999. Architektur und Ökologie – Eine Studie über die in der Baubranche vorhandene Meinung zu umweltgerechten Bauweisen. ETH Zürich, Laboratorium für technische Chemie, Professur für Sicherheit und Umweltschutz in der Chemie, Universitätsstrasse 31, 8092 Zürich, 1999.

Lehmann S, Nachhaltigkeit SIA. 1999. S. In: Lalive d'Epinay A, Quack D (Hrsg.) Ökologische Bewertung von Gebäuden zwischen Forschung und Praxis. Begleitende Unterlagen zum 11. Diskussionsforum Ökobilanzen vom 1. November 1999, ETH Zürich, Gruppe für Sicherheit und Umweltschutz, Laboratorium für Technische Chemie, Zürich, Oktober, 1999.

Le Téno JF. Développement d'un modéle d'aide à l'évaluation et à l'amélioration de la qualité environnementale des produits de construction. Genie civil et sciences de l'habitat, Université de Savoie, 212 Seiten, Savoie, 1996.

Le Téno JF. Use of LCA for the Measurement of Building Products Environmental Quality, IEA Task 18, A4 - Environmental Impact, CSTB, Grenoble 1997.

Levin H. Forum for a Healthy Built Environment. Brisbane, July, 1997.

Life-Cycle Energy Use in Office Buildings. Aug 94. Prepared by the Environmental Research Group, School of Architecture, University of British Columbia.

Lim R, en Lindeijer E et al. Milieubeoordeling van beglazingssystemen: Eindrapportage. (glazing), Amsterdam : IVAM Environmental Research, 1994, IVAM rapport; 94-03.

Lünser H. Ökobilanzen im Brückenbau: eine umweltbezogene, ganzheitliche Bewertung. Birkhäuser Verlag Zürich, 1999.

Maibach M, Peter D, Seiler B. Ökoinventar Transporte, Technischer Schlussbericht, Auftrag No 5001-34730, ISBN 3-9520824-5-7, infras, Zürich, 12 (1995).

Meil JK, Trusty WB. Life-Cycle Analysis of Building Materials and Impact Indicators. Proceedings: RILEM Workshop on Environmental Aspects of Building Materials and Structures, Espoo, 1995. Finland. 10 pp.

Meil JK, Trusty WB. Life-Cycle Assessment of Building Materials: An Appraisal Proceedings: Forest Products Society 1995 Annual Meeting, Portland, Oregon. June 1995. 10 pp.

Meil JK, Trusty WB. Wood Products LCA in Canada - A Review. Proceedings: 4th Eurowood Symposium. Stockholm Sweden, September 1997. 3 pp.

Meil JK. Building Materials in the Context of Sustainable Development: An Overview of the Research Program and Model. Life-Cycle Analysis (LCA) - a Challenge for Forestry and Forest Products: Proceedings of an International Workshop sponsored by the European Union, the European Forest Institute and the University of Hamburg. Hamburg, Germany May 1995. 18 pp.

Milieubeoordeling oeverbeschoeiingsmaterialen : Eindrapportage. (materials for shores/banks) DHV AIB BV en RIZA, 1997, N0333.01.001.

Milieuvergelijking tussenwanden in kantoorgebouwen. (interior separation walls) DHV facilities BV, Nationaal Onderzoekprogramma Hergebruik van afvalstoffen (NOH), 1996, 9623.

Nijdam DS. en Korenromp RHJ. (i.o.v. RIZA) Deventer: Tauw Milieu BV, Nadere analyse LCA zinken dakgoten. (zinc gutters), 1996, R3466248.DSN/M02.

Nordic Council of Ministers, Environmental data for building materials in the nordic countries. Nordic Council of Ministers, Kopenhagen, 1995.

Oorschot GF, van en Niemoller LG. Project. Verzamelen van Milieukengetallen van itumineuze dakbedekkingsmaterialen t.b.v. milieurelevante productinformatie (MrPi). (bituminous roofing material). Zeist: Stichting Dak en Milieu, 1998, GO/LN/98001.

Ospelt Ch: Der direkte und der indirekte Energieverbrauch der Haushalte in der Schweiz - Konzept zur Berechnung unter Verwendung der Input-Output-Analyse, Semesterarbeit in der Gruppe für Energieanalysen ETHZ, Zürich 1995.

Paehlke R. Ecological Carrying Capacity Effects of Building Materials Extraction. Sept 93. Environmental Policy Research. (includes separately bound annotated bibliography), Athena Sustainable Materials Institute Reports.

Paulsen J. Life Cycle Assessment for Building Products with Special Focus on Maintenance and Impacts from Usage Phase. Licentiate of Engineering Thesis. KTH, 1999.

Peuportier B. The life cycle simulation method EQUER applied to building components, CIB Conference : Construction and the environment, Gävle (Sweden), June 1998.

Polster B. Contribution à l'étude de l'impact environnemental des bâtiments par analyse du cycle de vie, thèse de doctorat, Ecole des Mines de Paris, 1995, 268 p. Regener project, European methodology for the evaluation of environmental impacts of buildings -life cycle assessment-, European Commission, DG XII for Science, Research and Development, Programme APAS, Contract RENA-CT940033, 1997.

Pulli R. 1998. Überblick über die Ökobilanzierung von Gebäuden. Beitrag im Rahmen des nationalen Projektes IEA CBS Annex 31. ETH Zürich, Professur für Sicherheit und Umweltschutz in der Chemie, Universitätsstrasse 31, 8092 Zürich, 1998.

Quack D. 1997. Komplexe Zusammenhänge, Ökobilanzen von Gebäuden liefern wichtige Informationen für die Bauplanung. In: Müllmagazin, Volume 2, Seiten 11–14. RHOMBOS-Verlag, 1997.

Quack D. 1999. Einfluss von Energiestandard und konstruktiven Faktoren auf die Umwelt- auswirkungen von Wohngebäuden anhand des Demonstrationsprojekts Niedrigenergiehäuser Heidenheim – eine Ökobilanz. Dissertation an der rheinisch-westfälischen Universität Aachen, Aachen, im Druck, 1999.

Rademaker, Luuk, Eindhoven: Technische Univer- siteit Eindhoven, Faculteit Bouwkunde, Milieueffecten van mechanisch bevestigde bitumineuze dakbanen: Afstudeerverslag. (bituminous roofing material), 1995, Rapportnummer; 9512 m.

Richter K. Ökobilanz von Fenstern, in: Fassade, Schweizerische Metallunion, S. 17–23, Zürich 1996.

Richter K, Brunner K, Bertschinger H. 1996. Ökologische Bewertung von Wärmeschutzgläsern, Integraler Vergleich verschiedener Verglasungsvarianten. Eidg. Materialprüfungs- und Forschungsanstalt (EMPA), Dübendorf.

Richter K, Fischer M, Gahlmann H, Menard M. 1995. Energie- und Stoffbilanzen bei der Herstellung von Wärmedämmstoffen. Eidg. Materialprüfungs- und Forschungsanstalt (EMPA), Dübendorf.

Richter K, Künniger T, Brunner K. 1996. Ökologische Bewertung von Fenster- konstruktionen verschiedener Rahmen- materilien (ohne Verglasung). EMPA-SZFF- Forschungsbericht, Schweizerische Zentral-

stelle für Fenster- und Fassadenbau (SZFF), Dietikon.

Redle M. 1999. Kies- und Energiehaushalt urbaner Regionen in Abhänigkeit der Siedlungsentwicklung. Dissertation Nr. 13108 der Eidgenössischen Technischen Hochschule Zürich, 1999.

Rijk AP et al. Vernieuwbare plaatmaterialen in de woningbouw: Milieu-effecten en toepasssing van plaatmaterialen van vernieuwbare grondstoffen in het interieur. (plate material for interior applications), Guda: WoonEnergie, 1995, DGM projectnummer; 93221414, WIE projectnummer; 583.

Roorda AAH, en Ven BL van der. Bladlood en het milieu. (Building lead), Apeldoorn : TNO Milieu, Energie en Procesinnovatie, 1999, TNO-MEP-R 98/503a.

Roskamp H, en Hoefnagels F. Drie bestemmingen voor fosfogips: een LCA leerproject. (phosophogypsum), Amsterdam: CREM, 1994.

Schmidt M. Environmentally effective building materials technology. in IABSE, 15th Con- gress. Vol. I, Seiten: 363 - 375, International Association for Bridge and Structural Engi- neering, Kopenhagen, 1996.

Schuurmans A. (INTRON), Siemens, A.J.M. en Vrouwenvelder, A.C.W.M. (TNO Bouw), Milieumatenconcept voor de bouw probabilistische uitwerking proefproject weg. (motorway road), 1994, CUR.

Schuurmans A. Milieuprofiel en milieumaten van een betonnen buitenriolering: Intron rapport nummer 95027 (concrete sewer pipes), Rioleringstechniek, 6e jrg, nr. 2, september 1995.

Schuurmans A. INTRON, Milieuprofiel van Trespa G2 plaatmateriaal, December 1992. (TRESPA cladding panels), TRESPA Hoechst Holland N.V, G11540, INTRON, September 1993.

Seijdel RR. Milieuprofielen van isolatieprodukten in hun toepassing. (insulation), Amersfoort: DHV AIB BV, 1995, K0748.01.001, JBJ/RS/MTB-1235.

Seijdel RR. European LCA-Data for expanded polystyrene building products. Bodegraven: PRC Bouwcentrum, 1988, 886.001.

SIA D0137. 1996. Stofer B, Bürgi H, Stocker M, Mercier C, Hodel N, Morandini G, Zingg M. Checklisten für energiegerechtes, ökologisches Bauen. SIA D0137, Schweizer Ingenieur und Architekt, Zürich, 1996.

SIA D0152. 1998. Koch P, Seiler B, Ott W, Lalive d'Epinay A, Gilgen D, Gugerli H. Instrumente für ökologisches Bauen im Vergleich. Ein Leitfaden für das Planungsteam. Beitrag im Rahmen des nationalen Projektes IEA CBS Annex 31. SIA D0152. Schweizerischer Ingenieur- und Architekten-Verein, Zürich, 1998.

Sidoroff S. Batimpact HQE. Vortragsunterlagen ATEQUE, Paris, 27. 6, 1996.

Steinle P, Lalive d'Epinay A.1999. Bau-Umweltbelastungskennwerte zur Abschätzung der Umwelt-verträglichkeit von Gebäuden in frühen Planungsphasen. ETH Zürich, Gruppe für Sicherheit und Umweltschutz, Laboratorium für Technische Chemie, Universitätsstrasse 31, Zürich, November 1999.

Tellenbach M. Abfallverbrennung in Zementwerken: Sinnvolle Verwertung oder gefährliche Konkurrenz für die Entsorgungsanlagen? in BUWAL, Abfallwirtschaft. Vol. 4/95, BUWAL-Bulletin, BUWAL, Bern, 1995.

The Summary Reports: Phase II and Phase III, Oct./93 and Jan./95 Prepared by Forintek Canada Corp. and Wayne B. Trusty and Associates. (includes 1997 version of The Research Guidelines), Athena Sustainable Materials Institute Reports.

Tillman A-M et al. Choice of system boundaries in life cycle assessment, in : Journal of Cleaner Production,1994 Vol. 2 Number 1 pp 21–29, 1994.

Todd JA. Scientific Consulting Group. 1997. Streamlined Environmental Lifecycle Assessment: An Approach for Evaluating the Environmental Performance of Building Materials. Sponsor: Environmental Protection Agency, Research Triangle Park, NC Air Pollution Prevention and Control Div; American Inst. of Architects, Washington, DC.

Trusty WB. 'Assessing the Ecological Carrying Capacity Effects of Resource Extraction' Life-Cycle Analysis (LCA) - a Challenge for Forestry and Forest Products: Proceedings of an International Workshop sponsored by the European Union, the European Forest Institute and the University of Hamburg. Hamburg, Germany May 1995. 10 pp.

Trusty WB, Meil JK. 'ATHENA': An LCA Decision Support Tool for the Building Community' Proceedings: Eco-indicators for Products and Materials - the State of Play 1997. Hosted by CANMET, Toronto, Ontario, November 25, 1997 8 pp.

Trusty WB, Meil JK. 'ATHENA™: An LCA Decision Support Tool - Application, Results and Issues' Proceedings: Second International Conference on Buildings and the Environment. Sponsored by CSTB and CIB TG8, Paris, France, June 1997 12 pp.

Trusty WB, Meil JK. 'ATHENA™: An LCA Model for the Building Design and Research Communities' Proceedings: 89th Annual Meeting of the Air and Waste Management Association, Nashville, Tennessee, June 1996 13 pp.

Trusty WB, Meil JK. 'Sustainable Buildings: The ATHENA™ Project Approach' Presented at the Sustainable Use of Materials international seminar sponsored by BRE in association with RILEM, Garston Watford, U.K. Sept. 1996 11 pp.

Trusty and Associates, dr R Paehlke. Environmental Policy Research, Assessing the Relative Ecological Carrying Capacity Impacts of Resource Extraction. Aug 94. Athena Sustainable Materials Institute Reports.

Trusty and Associates, dr R Paehlke. Environmental Policy Research, The Ecological Effects of Resource Extraction in Ontario. Mar 97, Athena Sustainable Materials Institute Reports.

Trusty WB, Meil JK, Venta G. 'Life Cycle Analysis of Structural Concrete and Gypsum Building Products - ATHENA™ Building Materials Project' Proceedings: IBAUSIL, 13th International Conference on Building Materials, Weimar, Germany, Sept. 1997 11 pp.

Venta G. 'Sustainable Building Materials: Use of Coproducts and Wastes in Gypsum and Concrete Bonded Boards' Proceedings: 5th International Inorganic Bonded Wood and Fiber Composite Materials Conference, Spokane, USA, Sept. 1996 12 pp.

Venta G, Trusty WB, Meil JK. 'ATHENA™ Life Cycle Analysis of Gypsum Wallboard and Associated Products' Proceedings: 5th International Conference on FGD and Synthetic Gypsum, Toronto, CA, May 1997 8 pp.

Versteeg H, Helm P, van der en Broers J. Checklist materialen en Milieu: Materiaalkeuze voor de wegenbouw, gericht op duurzaam bouwen.) Road en waterworks materials), Delft: Rijkswaterstaat, Dienst Weg- en Waterbouwkunde, RWW 19956, Rapport; IG-R-95185.

von Arx U. 1995. Bauprodukte und Zusatzstoffe in der Schweiz. Schriftenreihe Umwelt No. Nr. 245, BUWAL, Bern, 1995.

von Däniken A Chudacoff M et al. 1992. Vergleichende ökologische Bewertung von Anstrichstoffen im Baubereich, Daten. Bundesamt für Umwelt, Wald und Landschaft, Bern, 1992.

Vroegop, ir. MP. (shells), Isoschelp. LCA isolatieschelpen. Eindrapportage. L1327.01.001. Kenmerk RS/MVr/MTB-3148, DHV AIB B.V, vestiging Amersfoort, 1996.

Werner F, Richter K. 1997. Ökologische Untersuchung von Parkettfussböden, Betrachtung von Mosaik-Klebeparkett, Fertigparkett, 2-schichtig und Fertigparkett 3-schichtig. EMPA/ISP-Forschungsbericht, Dübendorf, Heimberg.

Werner F, Richter K, Bosshart S, Frischknecht R. 1997. Ökologischer Vergleich von Innenbauteilen am Beispiel von Zargen aus Massivholz, Holzwerkstoff und Stahl. Bericht EMPA und ETH, Dübendorf und Zürich, März, 1997.

Windsperger A, Steinlechner S, Piringer M. 1998. KURZFASSUNG. Ökologische Betrachtung von Fensterrahmen aus verschiedenen Werkstoffen. Forschungsinstitut für Chemie und Umwelt - TU Wien. Vertreter von Herstellern der einzelnen Produkte. (im Auftrag der Niederösterreichischen Landesregierung).

Wirtz, W, Dämmstoffe: Entwarnung für Mineralfasern. natur: 96 - 103, 1996.

Working with ATHENA: Comparative Manual and Model Case Study Assessments. Mar 97. Prepared by the Environmental Research Group, School of Architecture, University of British Columbia. Athena Sustainable Materials Institute Reports.

Zellweger C, Hill M, Gehrig R, Hofer P. 1997. Schadstoffemissionsverhalten von Baustoffen. Forschungsprogramm Rationelle Energienutzung in Gebäuden No. 2. Auflage, EMPA, Dübendorf, January, 1997.

Product Life-Cycle Studies

[AfB] Amt für Bundesbauten. 1995. Nutzungszeiten von Gebäuden und Bauteilen. AfB, Bern, January 1995.

Baitz M, Kreissig J, Betz M, Eyerer P, Kümmel J, Reinhardt H. 1997. Life Cycle Assessment of Buildings, Materials and Structures in Germany. University of Stuttgart.

Baumann Th, Itschner L. 1998. Umweltrelevanz der Haustechnik. Eine Entscheidungsgrundlage. Beitrag im Rahmen des nationalen Projektes IEA CBS Annex 31. ETH Zürich, Professur für Sicherheit und Umweltschutz in der Chemie, Universitäts-strasse 31, 8092 Zürich, 1998. ISBN 3-906734-04-8.

Björklund T, Jönsson A, Tillman AM. 1996. LCA of Building Frame Structures. Environmental Impact over the Life Cycle of Concrete and Steel Frames. Report 1997:2 Technical Environmental Planning. Göteborg, Sweden.

Björklund T, Tillman AM. 1997. LCA of Building Frame Structures. Environmental Impact over the Life Cycle of Wooden and Concrete Frames. Report 1997:2 Technical Environmental Planning. Göteborg, Sweden.

Building Assemblies: Construction Energy & Emissions. Aug 93. Prepared by the Environmental Research Group, School of Architecture, University of British Columbia, ATHENA Sustainable Materials Institute Reports (Canada).

CANMET & Radian Canada Inc. Raw Material Balances, Energy Profiles and Environmental Unit Factor Estimates for Cement and Structural Concrete Products. Aug 93. ATHENA Sustainable Materials Institute Reports (Canada).

Caspersen N. Accumulated energy consumption and emissions of CO_2, SO_2, and NO_x during life cycle of stainless steel. Ironmaking and Steelmaking 23, (4):Págs.317-323 (1996). DQFUAB. ABO159.

Chevalier J, Le Téno J. Life Cycle Analysis with Ill-Defined Data and its Application to Building products. Int J LCA 1, (2):Págs.90-96 (1996). 0949-3349.

Cretton P. Influence du plan quartier sur les reseaux et les impacts environnementaux, Master europeen en architecture et developpement durable, Institut de technique du bâtiment de l'EPFL, Lausanne 1997.

Dinesen, J, Traberg-Borup, S.: An Energy Life Cycle Assessment Model for Building Design, Danish Building Research Institute SBI, Denmark 1994.

Doka G. 2000. Ökoinventar der Entsorgungsprozesse von Baumaterialien – Grundlagen zur Integration der Entsorgung in Ökobilanzen von Gebäuden. Beitrag im Rahmen des Projektes IEA CBS Annex 31, Laboratorium für Technische Chemie, Gruppe für Sicherheit und Umweltschutz, Zürich. zu beziehen bei: ZEN, EMPA Dübendorf, 2000.

Forintek Canada Corp. Raw Material Balances, Energy Profiles and Environmental Unit Factor Estimates for Structural Wood Products. Mar 93. Athena Sustainable Materials Institute Reports.

Frischknecht R, Bollens U, Bosshart St, Ciot M, Ciseri L, Doka G, Dones R, Gantner U, Hischier R, Martin A. 1996. Ökoinventar von Energiesystemen, Grundlagen für den ökologischen Vergleich von Energiesystemen und den Einbezug von Energiesystemen in Ökobilanzen für die Schweiz, Institut für Energietechnik, Laboratorium für Energiesysteme, Gruppe Energie-Stoffe-Umwelt, 3. Auflage, ENET, Bern, 1996.

Frischknecht R, Hofstetter P, Knoepfel I, Meenard M. 1991. (Redaktion), Ökoinventare für Energiesysteme. Institut für Energietechnik, Laboratorium für Energiesysteme, Gruppe Energie-Stoffe-Umwelt, ETH Zürich, 1. Auflage, ENET, Bern, 1991.

Gordon M. Engineering Demolition Energy Analysis of Office Building Structural Systems. Mar 97. ATHENA Sustainable Materials Institute Reports (Canada).

IP Bau. 1991. Recycling - Verwertung und Behandlung von Bauabfällen. Impulsprogramm Bau, Bundesamt für Konjunkturfragen, Bern, 1991.

Norris GA, Sylvatica. Life Cycle Inventory Analyses of Building Envelope Materials: Update and Expansion. Jun 99. ATHENA Sustainable Materials Institute Reports (Canada).

Hebel AG: Die 1. Ökobilanz für ein Haus - Unser Baustoff im Gefüge der Umwelt, Hebel Mitteilung, Fürstenfeldbruck 1995.

Meil JK. A Life Cycle Analysis of Solid Wood and Steel Cladding. Sep 98. ATHENA Sustainable Materials Institute Reports (Canada).

Meil JK. LCA of Solid Wood and Steel Cladding . Sep 98. ATHENA Sustainable Materials Institute Reports (Canada).

Kasser U, Pöll M. 1995. Graue Energie von Baustoffen. 1. Auflage, Büro für Umweltchemie, Zürich, 1995.

Kasser U, Pöll M. 1998. Graue Energie von Baustoffen. 2. Auflage, Büro für Umweltchemie, Zürich, 1998.

Kreissig et al. LCA of building materials, constructions and total buildings. Research project prepared by IKP (Institute for Polymer Testing and Polymer Science) and IWB (Institute for Building Materials) of the University of Stuttgart in co-operation with 45 German building material producers or associations. Finished in 1999.

Kreissig et al. LCA of windows and facades. 1998. Prepared by IKP (Institute for Polymer Testing and Polymer Science) and IWB (Institute for Building Materials) of the University of Stuttgart in co-operation with the German window and facade association.

Maries A, Parrott L. 'Environmental analysis case study of concrete road construction options' Report for CIA/DETR project Defining and improving environmental performance in the concrete industry, December 1999, 9 pages.

Maibach M, Mauch S, Iten R, Banfi S, Ott W, Ledergerber E, Masuhr K. 1996. Die vergessenen Milliarden, Externe Kosten im Energie- und Verkehrsbereich. Infras, Haupt Verlag, Bern, 1996.

Norris GA. An Exploratory Life Cycle Study of Selected Building Envelope Materials. Jul 98. Sylvatica, ATHENA Sustainable Materials Institute REPORTS (Canada).

Ospelt Ch.: Ökobilanz von Gebäuden: Methodik und Anwendung - Unter besonderer Betrachtung von Rohbau und Heizungsanlage, EPFL Lausanne, Lausanne 1995.

Parrott L. 'Effect of changes in cement manufacture on environmental performance of concrete' Environmental analysis case study for CIA/DETR project Defining and improving environmental performance in the concrete industry, December 1999, 6 pages.

Parrott L. 'Environmental report for the UK concrete industry 1994 to 1998' Report for CIA/DETR project Defining and improving environmental performance in the concrete industry, November 1999, 8 pages.

Parrott LJ. 'Recycled concrete as aggregate in new concrete' CIA/DETR project Defining and improving environmental performance in the concrete industry, Environmental factsheet, September 1999, 7 pages.

Parrott LJ. 'Environmental impacts of transport relative to those of concrete' CIA/DETR project Defining and improving environmental performance in the concrete industry, Environmental factsheet, November 1999, 6 pages.

Parrott L, Cleverly C. 'Effect of formwork on the environmental impact of concrete' CIA/DETR project Defining and improving environmental performance in the concrete industry, Environmental factsheet, November 1999, 5 pages.

Parrott L, Higgins D, Sear L. 'Pulverised-fuel ash and ground granulated blastfurnace slag environmental analysis case study' Report for CIA/DETR project Defining and improving environmental performance in the concrete industry, August 1999, 6 pages.

Richter K, Fischer M, Gahlmann H. 1995. Hrsg.: Bundesamt für Energiewirtschaft. Energie- und Stoffbilanzen bei der Herstellung von Wärmedämmstoffen. Abschlussbericht über das BEW-Forschungsprojekt Heft Forschungsprogramm Rationelle Energienutzung in Gebäuden. Bern, 1995.

Schneeberger K. 1999. Kunststoffe in der Schweizer Bauindustrie. Eine Studie im Rahmen des IEA BCS Annex 31, durchgeführt bei der DOW Europe S.A, Horgen, ETH Zürich, Laboratorium für Technische Chemie, Gruppe für Sicherheit und Umweltschutz, Universitätsstrasse 31, 8092 Zürich, 1999.

SIA 493. 1997. Schweizer Ingenieur- und Architekten-Verein, Deklaration ökologischer Merkmale von Baustoffen. SIA Norm 493. Zürich, 1997.

SIA D046. 1989. Schweizer Ingenieur- und Architekten-Verein, Schadstoffarmes Bauen. SIA Dokumentation D046, Zürich, 1989.

SIA D0118. 1993. Schweizer Ingenieur- und Architekten-Verein, Ökologie in der Haustechnik – eine Orientierungshilfe. SIA Dokumentation D0118, Zürich, 1993.

SIA D0123. 1995. Intep AG, Steiger P. Hochbaukonstruktionen nach ökologischen Gesichtspunkten. SIA-Dokumentation D0123, Schweizer Ingenieur- und Architekten-Verein, Zürich, September, 1995

Stahel U, Fecker I, Förster R, Maillefer C, Reusser L. 1998. Bewertung von Ökoinventaren für Verpackungen. Schriftenreihe Umwelt Nr. 300, Bundesamt für Umwelt, Wald und Landschaft, Bern, 1998.

Steinle P, Lalive d'Epinay A. 2000. Bau-Umweltbelastungskennwerte zur Abschätzung der Umweltverträglichkeit von Gebäuden in frühen Planungsphasen. ETH Zürich, Gruppe für Sicherheit und Umweltschutz, Laboratorium für Technische Chemie, Universitätsstrasse 31, 8092 Zürich, 1999.

Stelco Technical Services Limited. Raw Material Balances, Energy Profiles and Environmental Unit Factor Estimates for Structural Steel Products. Aug 93. ATHENA Sustainable Materials Institute REPORTS (Canada).

Steltech. Raw Material Balances, Energy Profiles and Environmental Unit Factor Estimates for Mini-Mill Steel Materials. Aug 94. ATHENA Sustainable Materials Institute REPORTS (Canada).

Todd JA. Streamlined Environmental Lifecycle Assessment: An Approach for Evaluating the Environmental Performance of Building Materials. (Final rept. 1992-97.), Scientific Consulting Group, Inc, Gaithersburg, Sponsor: Environmental Protection Agency, Research Triangle Park, NC. Air Pollution Prevention and Control Div; American Inst. Of Architects, Washington, DC.

Tschirner B. Démarche pratique pour l'établissement de l'ecobilan d'un bâtiment, Diplomarbeit an der EPF Lausanne, Lausanne 1995.

Venta GJ, Glaser & Associates. Life Cycle Analysis of Brick and Mortar Products. Sep 98, ATHENA Sustainable Materials Institute Reports (Canada).

Venta, Glaser and Associates. Life Cycle Analysis of Gypsum Wallboard and Associated Finishing Products. Mar 97. ATHENA Sustainable Materials Institute REPORTS (Canada).

Weibel Th, Stritz A. 1995. Ökoinventare und Wirkungsbilanzen von Baumaterialien. ESU-Reihe No. 1/95, ETH, Zürich. zu beziehen bei ENET, Bern. September, 1995.

Websites

- annex31.wiwi.uni-karlsruhe.de/
- Greenbuilding.ca/gbc98.html#MEMBERS
- www. ikpgabi.uni-stuttgart.de
- www.aloha.net/~laumana
- www.aloha.net/~laumana/index.html
- www.athenaSMI.ca (software BEES)
- www.bioarchitettura.org
- www.bre.co.uk/envprofiles
- www.ebuild.com
- www.ekologik.cit.chalmers.se/
- www.gabi-software.com
- www.greenbuilder.com
- www.ivambv.uva.nl
- www.esu-services.ch
- www.ogip.ch
- www.empa.ch/zen
- www.trentu.ca/faculty
- www.unite.ch/doka/lca.htm
- www.uni-weimar.de/SCC/PRO
- www.uns.umnw.ethz.ch/uns/ (page Discussion Forum on LCA, link on the left side)

Software and Databases

Building specific

- ATHENA 1.0
- BEES 1.0 (software)
- Danish Building Research Institute database (uses the Danish EDIP method to normalize environmental effects)
- ECO-Quantum
- ENVEST www.bre.co.uk/envest
- GaBi 3, software
- GBA-Tool version 1.21
- LCA HOUSE, software including databases http://www.vtt.fi/rte/esitteet/ymparisto/lcahouse.html
- OGIP (software)

General

- IDEMAT database
- IKP database module for building materials additional to the software and database GaBi 3 professional
- IVAM LCA Data 2.0 Database for building materiales in Holland (database in SimaPro 4 format)
- LESOSAI4
- LcaiT 4.0 (software)
- SimaPro 4.0 (software)
- Swiss Database ECOINVENT, ETH Zürich, Prof. K. Hungerbühler, Gruppe für Sicherheit und Umweltschutz; EMPA Dübendorf, ZEN, Hr. M. Zimmermann; esu-services, Dr. R. Frischknecht.
- Merkmale von Baustoffen. SIA Norm 493. Zürich, 1997.

Abbreviations and Definitions

(A specific LCA definition list can be found in the ISO14040 series)

Additives: Materials/products added in small (mass) quantities to affect the properties of the final product or material.

Admixture: Additive added during mixing.

Allocation: Partitioning the input or output flows of a process to the product system under study.

B/C: Building and construction

BMCCs: Building Material and Component Combinations. A building material, building product or building component or a combination of these.

Building: Construction works that has the provision of shelter for its occupants or contents as one of its main purposes and is normally designed to stand permanently in one place. (ISO 6707-1:1989)

Building component: Product manufactured as a distinct unit to serve a specific function or functions (ISO 6707-1: 1989, ISO 15686-1).

Building element: Major functional part of a building (ISO 6707-1: 1989). Examples: foundation, floor, roof, wall, services.

Building material: Substance that can be used to form building products or construction works (ISO 6707-1: 1989, prEN 1745, ISO 15686-1). Examples: cement, concrete mortar, wood, plastic granulates, etc. Materials are usually non-designed products or semi-manufactures.

Building product: Item manufactured or processed for incorporation in the construction works (ISO 6707-1: 1989, ISO 15686-1).

CEPMC: Council of European Producers of Materials for Construction

CIB: International Council for Research and Innovation in Building and Construction

CML: Centrum voor Milieukunde Construction

Construction: Assembled or complete part resulting from work on site (ISO6707-1: 1989).

Construction work: Activities of forming a construction works or construction (ISO 6707-1: 1989).

Construction works: Everything that is constructed or results from construction operations (ISO 6707-1: 1989).

EDIP: Environmental development of industrial products

EIA: Environmental impact assessment

EQUER: Evaluation de la Qualité Environnementale des Bâtiment

ETH: Eidgenössische Technische Hochschule

Functional unit: A reference unit for LCA, which describes either the function or functions of a building or construction, or of a building component or product used in a building or construction. The description is in terms of the quantitative performance, which must be achieved within a specific service life. (Free definition on behalf of this report) (In accordance with the Building Product Directive: The products must be suitable for use in construction works, which are fit for their intended use…and in this connection satisfy the following [performance] require-ments…. Such requirements must, subject to normal maintenance, be satisfied for an economically reasonable working life.) A **functional unit on the BMCC-level** often does not contain a final function in a building or construction (e.g., in case of building materials), but describes the BMCC, contains a quantity and possibly also indicates the application and intended service life of application. Also called 'per kg' unit or 'analysis unit'. Examples: 1 m³ ready mix concrete, 1 piece of brick, 1 m² load-bearing wall. Although services and activities during the life cycle of a building are not BMCCs, they can be treated as such. Example: 'demolition of 10t of concrete'.

IEA: International Energy Agency

ISO: International Organization for Standardization

KEMI: Kemikalieinspektionen

LCA: Life-cycle assessment

LCC: Life-cycle costing

LCI: Life-cycle inventory

LCIA: Life-cycle impact assessment

NVTB: Nederlands Verbong Toelevering Bouw

OGIP: Optimierung der Gesamtanforderungen für die Integrale Planung

Performance: Behaviour related to use (ISO 6707-1: 1989). Ability of a building or its parts to fulfil the required functions under the intended use conditions (CIB W60).

Product chain: The set of consecutive links in the production of a product. Together, these links form a chain.

Product group: A group of building materials, building products, or building components that all have the same main application or function.

Product system: Collection of materially and energetically connected unit-processes, which perform one or more defined functions

Raw material: A substance extracted from the environment to manufacture a material or product

RICS: Royal Institute of Chartered Surveyors

SBR: Stichting Bouwresearch

Service life: Period of time after installation during which a building or its parts meet the exceed the performance requirements (ISO 15686-1)

SETAC: Society of Environmental Toxicology and Chemistry

SFT: Statens Forunensningstilsyn

Type I environmental declaration: Environmental labelling; definition in ISO 14024:1999.

Type III environmental declaration: Quantified environmental data for a product with pre-set categories of parameters based on ISO 14040-series of standards, but not excluding additional environmen-tal information provided within a Type III environmental declaration program (ISO/TR 14025:2000).

Type III environmental declaration program: Voluntary process by which an industrial sector or independent body develops a Type III environmental declaration, including setting minimum require-ments, selecting categories of paramaters, defining the involvement of third parties and the format of external communications.

Unit-process: Smallest portion of a product system for which data are collected when a life-cycle assessment is performed

User: Organisation, person, animal, or object for which a building is designed (ISO 6707-1:1989)

VTT: Valtion Teknillinen Tutkimuskeskus (Technical Research Centre of Finland)

Index

SETAC

A Professional Society for Environmental Scientists and Engineers and Related Disciplines Concerned with Environmental Quality

The Society of Environmental Toxicology and Chemistry (SETAC), with offices currently in North America and Europe, is a nonprofit, professional society established to provide a forum for individuals and institutions engaged in the study of environmental problems, management and regulation of natural resources, education, research and development, and manufacturing and distribution.

Specific goals of the society are:
- Promote research, education, and training in the environmental sciences.
- Promote the systematic application of all relevant scientific disciplines to the evaluation of chemical hazards.
- Participate in the scientific interpretation of issues concerned with hazard assessment and risk analysis.
- Support the development of ecologically acceptable practices and principles.
- Provide a forum (meetings and publications) for communication among professionals in government, business, academia, and other segments of society involved in the use, protection, and management of our environment.

These goals are pursued through the conduct of numerous activities, which include:
- Hold annual meetings with study and workshop sessions, platform and poster papers, and achievement and merit awards.
- Sponsor a monthly scientific journal, a newsletter, and special technical publications.
- Provide funds for education and training through the SETAC Scholarship/Fellowship Program.
- Organize and sponsor chapters to provide a forum for the presentation of scientific data and for the interchange and study of information about local concerns.
- Provide advice and counsel to technical and nontechnical persons through a number of standing and ad hoc committees.

SETAC membership currently is composed of more than 5,000 individuals from government, academia, business, and public-interest groups with technical backgrounds inchemistry, toxicology, biology, ecology, atmospheric sciences, health sciences, earth sciences, and engineering.

If you have training in these or related disciplines and are engaged in the study, use, or management of environmental resources, SETAC can fulfill your professional affiliation needs.

All members receive a newsletter highlighting environmental topics and SETAC activities, and reduced fees for the Annual Meeting and SETAC special publications.

All members except Students and Senior Active Members receive monthly issues of *Environmental Toxicology and Chemistry (ET&C)*, a peer-reviewed journal of the Society. Student and Senior Active Members may subscribe to the journal. Members may hold office and, with the Emeritus Members, constitute the voting membership.

If you desire further information, contact the appropriate SETAC office.

SETAC North America
1010 North 12th Avenue
Pensacola, Florida 32501-3367 USA
T 850 469 1500   F 850 469 9778
E setac@setac.org

SETAC Europe
Avenue de la Toison d'Or 67
B-1060 Brussels, Belgium
T 32 2 772 72 81   F 32 2 770 53 83
E setac@setaceu.org

www.setac.org

Environmental Quality Through Science®

Other SETAC Books

Code of Life-Cycle Inventory Practice
de Beaufort-Langeveld, Bretz, van Hoof, Hischier, Jean, Tanner, Huijbregts
2003

Life-Cycle Impact Assessment: Striving Towards the Best Practice
Udo de Haes, Finnveden, Goedkoop, Hauschild, Hertwich, Hofstetter, Jolliet, Klöpffer, Krewitt, Lindeijer,
Müller-Wenk, Olsen, Pennington, Potting, Steen, editors
2002

Silver in the Environment: Transport, Fate, and Effects
Andren and Bober, editors
2002

Bioavailability of Metals in Terrestrial Ecosystems: Importance of Partitioning for Bioavailability in Invertebrates
Allen, editor
2002

Test Methods to Determine Hazards of Sparingly Soluble Metal Compounds in Soils
Fairbrother, Glazebrook, Van Straalen, Tararzona, editors
2002

Interconnections Between Human Health and Ecological Integrity
Di Giulio and Benson, editors
2002

Community-Level Aquatic System Studies-Interpretation Criteria
Giddings, Brock, Heger, Heimbach, Maund, Norman, Ratte, Schäfers, Streloke, editors
2002

Avian Effects Assessment: A Framework for Contaminants Studies
Hart, Balluff, Barfknecht, Chapman, Hawkes, Joermann, Leopold, Luttik, editors
2001

Impact of Low-Dose, High-Potency Herbicides on Nontarget
and Unintended Plant Species
Ferenc, editor
2001

Risk Management: Ecological Risk-Based Decision Making
Stahl, Bachman, Barton, Clark, deFur, Ells, Pittinger, Slimak, Wentsel, editors
2001

Ecotoxicology of Amphibians and Reptiles
Sparling, Linder, Bishop, editors
2000

Environmental Contaminants and Terrestrial Vertebrates:
Effects on Populations, Communities, and Ecosystems
Albers, Heinz, Ohlendorf, editors
2000

Evaluating and Communicating Subsistence Seafood Safety in a Cross-Cultural Context:
Lessons Learned from the Exxon Valdez *Oil Spill*
Field, Fall, Nighswander, Peacock, Varanasi, editors
2000

Multiple Stressors in Ecological Risk and Impact Assessment:
Approaches to Risk Estimation
Ferenc and Foran, editors
2000

Natural Remediation of Environmental Contaminants:
Its Role in Ecological Risk Assessment and Risk Management
Swindoll, Stahl, Ells, editors
2000

Endocrine Disruption in Invertebrates: Endocrinology, Testing and Assessment
DeFur, Crane, Ingersoll, Tattersfield, editors
1999

Linkage of Effects to Tissue Residues: Development of a Comprehensive Database for Aquatic Organisms Exposed to Inorganic and Organic Chemicals
Jarvinen and Ankley, editors
1999

Multiple Stressors in Ecological Risk and Impact Assessment
Foran and Ferenc, editors
1999

Reproductive and Developmental Effects of Contaminants in Oviparous Vertebrates
DiGiulio and Tillitt, editors
1999

Restoration of Lost Human Uses of the Environment
Grayson Cecil, editor
1999

Ecological Risk Assessment: A Meeting of Policy and Science
Peyster and Day, editors
1998

Ecological Risk Assessment Decision-Support System: A Conceptual Design
Reinert, Bartell, Biddinger, editors
1998

Principles and Processes for Evaluating Endocrine Disruption in Wildlife
Kendall, Dickerson, Geisy, Suk, editors
1998

Radiotelemetry Applications for Wildlife Toxicology Field Studies
Brewer and Fagerstone, editors
1998

Sixth LCA Symposium for Case Studies (Presentation summaries)
1998

Sustainable Environmental Management
Barnthouse, Biddinger, Cooper, Fava, Gillett, Holland, Yosie, editors
1998

Uncertainty Analysis in Ecological Risk Assessment
Warren-Hicks and Moore, editors
1998

Chemical Ranking and Scoring: Guidelines for Relative Assessments of Chemicals
Swanson and Socha, editors
1997

Chemically Induced Alterations in Functional Development and Reproduction of Fishes
Rolland, Gilbertson, Peterson, editors
1997

Ecological Risk Assessment for Contaminated Sediments
Ingersoll, Dillon, Biddinger, editors
1997

Fifth LCA Symposium for Case Studies (Presentation summaries)
1997

Life-Cycle Impact Assessment: The State-of-the-Art, 2nd ed.
Barnthouse, Fava, Humphreys, Hunt, Laibson, Moesoen, Owens, Todd,
Vigon, Weitz, Young, editors
1997

Public Policy Applications of Life-Cycle Assessment
Allen, Consoli, Davis, Fava, Warren, editors
1997

Quantitative Structure-Activity Relationships (QSAR) in Environment Sciences VII
Chen and Schüürmann, editors
1997

Simplifying LCA: Just a Cut?
Christiansen, editor
1997

Whole Effluent Toxicity Testing:
An Evaluation of Methods and Prediction of Receiving System Impacts
Grothe, Dickson, Reed-Judkins, editors
1996

Procedures for Assessing the Environmental Fate and Ecotoxicity of Pesticides
Mark Lynch, editor
1995

The Multi-Media Fate Model: A Vital Tool for Predicting the Fate of Chemicals
Cowan, D. Mackay, Feijtel, Meent, Di Guardo, Davies, N. Mackay, editors
1995

Aquatic Dialogue Group: Pesticide Risk Assessment and Mitigation
|Baker, Barefoot, Beasley, Burns, Caulkins, Clark, Feulner, Giesy, Graney,
Griggs, Jacoby, Laskowski, Maciorowski, Mihaich, Nelson, Parrish, Siefert, Solomon, van der Schalie, editors
1994

Integrating Impact Assessment into LCA
Udo de Haes, Jensen, Klepffer, Lindfors, editors
1994

Life-Cycle Assessment Data Quality: A Conceptual Framework
Fava, Jensen, Lindfors, Pomper, De Smet, Warren, Vigon, editors
1994

A Conceptual Framework for Life-Cycle Impact Assessment
Fava, Consoli, Denison, Dickson, Mohin, Vigon, editors
1993

Guidelines for Life-Cycle Assessment: A "Code of Practice"
Consoli, Allen, Boustead, Fava, Franklin, Jensen, De Oude, Parrish, Perriman, Postlethwaite, Quay, Séguin,
Vigon, editors
1993

A Technical Framework for Life-Cycle Assessment
Fava, Denison, Jones, Vigon, Curran, Selke, Barnum, editors
1991

Research Priorities in Environmental Risk Assessment
Fava, Adams, Larson, G. Dickson, K. Dickson, W. Bishop
1987